McGraw-Hill

Welcome to My Math — your very own math book!

You can write in it — in fact, you are encouraged to write, draw, circle, explain, and color as you explore the exciting world of mathematics. Let's get started. Grab a pencil and finish each sentence.

My name is _____.

My favorite color is _____.

My favorite hobby or sport is _____.

My favorite TV program or video game is

_____

My favorite class is _____.

Math, of course!

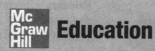

McGraw Hill Education

Bothell, WA • Chicago, IL • Columbus, OH • New York, NY

connectED.mcgraw-hill.com

 **Education**

STEM McGraw-Hill is committed to providing
instructional materials in Science, Technology, Engineering,
and Mathematics (STEM) that give all students a solid
foundation, one that prepares them for college and careers
in the 21st century.

Send all inquiries to:
McGraw-Hill Education
STEM Learning Solutions Center
8787 Orion Place
Columbus, OH 43240

ISBN: 978-0-02-115023-6 *(Volume 1)*
MHID: 0-02-115023-0

Printed in the United States of America.

13 14 15 16 QVS 18 17 16 15

Common Core State Standards© Copyright 2010.
National Governors Association Center for Best
Practices and Council of Chief State School Officers.
All rights reserved.

Our mission is to provide educational resources that enable
students to become the problem solvers of the 21st century
and inspire them to explore careers within Science, Technology,
Engineering, and Mathematics (STEM) related fields.

Meet The Artists!

Carolyn Phung

Invertebrate Art My class was studying invertebrates, and we discovered that insects have their own symmetry. Then we talked about radial symmetry which is like a wheel. I put these ideas together to form my artwork. *Volume 1*

Grace Kramer

Math is Awesome 4 Math challenges me to learn new things. When I think about math problems they are sometimes hard to solve just like puzzles. Math is puzzles that are all around me. *Volume 2*

Other Finalists

Jesus Pallares
Math Roller Coaster

Isidro Tavares
Righteous Math

Jaquise Hickman
Roadmap of Math

Avi Sanan
Jumping Fish

Isa Weiss
Sailing Through Math

Luis Rodriguez
Math is Awesome 12

Mikayla Pilgrim
My Many Colored Numbers

Kayley Spiller
The Part Tree

Carl Zent
Math Means a Working Community

Calista Smith
Skinny Jeans

Find out more about the winners and other finalists at www.MHEonline.com.

We wish to congratulate all of the entries in the 2011 *McGraw-Hill My Math* "What Math Means To Me" cover art contest. With over 2,400 entries and more than 20,000 community votes cast, the names mentioned above represent the two winners and ten finalists for this grade.

GO digital

it's all at
connectED.mcgraw-hill.com

Go to the Student Center for your eBook, Resources, Homework, and Messages.

Write your Username [_____] ✏ Password [_____] ✏

Get your resources online to help you in class and at home.

Vocab

Find activities for building vocabulary.

Watch

Watch animations of key concepts.

Tools

Explore concepts with virtual manipulatives.

Check

Self-assess your progress.

eHelp

Get targeted homework help.

Games

Reinforce with games and apps.

Tutor

See a teacher illustrate examples and problems.

GO mobile

Scan this QR code with your smart phone* or visit mheonline.com/stem_apps.

*May require quick response code reader app.

Available on the App Store

Contents in Brief
Organized by Domain

CCSS

Common Core State Standards

Standards for Mathematical PRACTICE 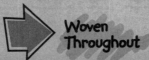 Woven Throughout

Chapter 1 Place Value

ESSENTIAL QUESTION
How does place value help represent the value of numbers?

Are you ready for the great outdoors?

Getting Started

Lessons and Homework

Wrap Up

Look for this!

Watch ▶

Click online and you can watch videos that will help you learn the lessons.

connectED.mcgraw-hill.com

Chapter 2
Add and Subtract Whole Numbers

Getting Started

Lessons and Homework

Wrap Up

connectED.mcgraw-hill.com

Chapter 3

Understand Multiplication and Division

ESSENTIAL QUESTION
How are multiplication and division related?

eHelp

Look for this!
Click online and you can get more help while doing your homework.

Chapter 4

Multiply with One-Digit Numbers

Getting Started

Lessons and Homework

Wrap Up

I hope I can multiply my savings!

connectED.mcgraw-hill.com

Chapter 5 Multiply with Two-Digit Numbers

(t)Digital Archive Japan/Alamy, (b)Sylvia Bors/The Image Bank/Getty Images
Copyright © The McGraw-Hill Companies, Inc.

ESSENTIAL QUESTION
How can I multiply by a two-digit number?

Getting Started

Lessons and Homework

Wrap Up

Tools
Look for this!
Click online and you can find tools that will help you explore concepts.

Chapter 6

Divide by a One-Digit Number

ESSENTIAL QUESTION
How does division affect numbers?

Getting Started

Lessons and Homework

Wrap Up

connectED.mcgraw-hill.com

Chapter 7
Patterns and Sequences

Operations and Algebraic Thinking

ESSENTIAL QUESTION
How are patterns used in mathematics?

Look for this!

Tutor

Click online and you can watch a teacher solving problems.

Chapter 8 Fractions

ESSENTIAL QUESTION
How can different fractions name the same amount?

Getting Started

Lessons and Homework

Wrap Up

Chapter 9
Operations with Fractions

Copyright © The McGraw-Hill Companies, Inc. (t)The McGraw-Hill Companies, (b)photosindia/Getty Images

ESSENTIAL QUESTION
How can I use operations to model real-world fractions?

Getting Started

Lessons and Homework

Wrap Up

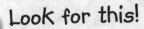

Vocab
σ_{b_c} **Look for this!**
Click online and you can find activities to help build your vocabulary.

Chapter 10

Fractions and Decimals

Getting Started

Lessons and Homework

Wrap Up

connectED.mcgraw-hill.com

Chapter 11 Customary Measurement

ESSENTIAL QUESTION
Why do we convert measurements?

Getting Started

Lessons and Homework

Wrap Up

Look for this!

Check ✓ Click online and you can check your progress.

Chapter 12
Metric Measurement

ESSENTIAL QUESTION
How can conversion of measurements help me solve real-world problems?

Chapter 13
Perimeter and Area

Chapter 14 Geometry

ESSENTIAL QUESTION
How are different ideas about geometry connected?

Getting Started

Lessons and Homework

Wrap Up

connectED.mcgraw-hill.com

Our Great Outdoors

ESSENTIAL QUESTION

How does place value help represent the value of numbers?

Watch

Watch a video!

MY Common Core State Standards

4.NBT.1 Recognize that in a multi-digit whole number, a digit in one place represents ten times what it represents in the place to its right.

4.NBT.2 Read and write multi-digit whole numbers using base-ten numerals, number names, and expanded form. Compare two multi-digit numbers based on meanings of the digits in each place, using >, =, and < symbols to record the results of comparisons.

4.NBT.3 Use place value understanding to round multi-digit whole numbers to any place.

Standards for
Mathematical
PRACTICE

Hey, I already know some of these!

1. Make sense of problems and persevere in solving them.
2. Reason abstractly and quantitatively.
3. Construct viable arguments and critique the reasoning of others.
4. Model with mathematics.
5. Use appropriate tools strategically.
6. Attend to precision.
7. Look for and make use of structure.
8. Look for and express regularity in repeated reasoning.

= focused on in this chapter

Name _Leo_

Am I Ready?

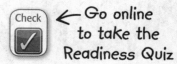

Check ✓ ← Go online to take the Readiness Quiz

Compare. Use >, <, or =.

1. 8,000 ⊘ 8,100   **2.** 3,404 ◯ 3,044   **3.** 7,635 ◯ 7,635

Round each number to the nearest ten.

4. 24
    20

5. 16
    20

6. 37
    40

Round each number to the nearest hundred.

7. 215
    200

8. 189
    200

9. 371
    400

10. A recipe calls for 11 eggs. Write this number in word form.

    eleven

Write each set of numbers in order from *least* to *greatest*.

11. 124, 139, 129
    124, 129, 139

12. 257, 184, 321
    184,

13. There are twenty-five students in Cooper's class. Write this number in standard form.

    _____

Shade the boxes to show the problems you answered correctly.

How Did I Do? → | 1 | 2 | 3 | 4 | 5 | 6 | 7 | 8 | 9 | 10 | 11 | 12 | 13 |

MY Math Words

Review Vocabulary

| hundreds | ones | ten thousands | tens | thousands |

Making Connections

Use the review vocabulary words to describe each digit in the diagram. Then answer the question.

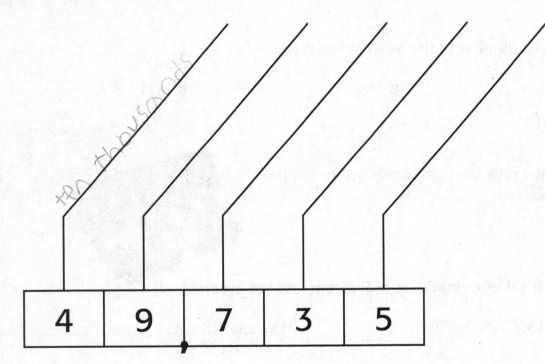

ten thousands

| 4 | 9 | 7 | 3 | 5 |

Suppose 49,735 people have tickets for a basketball game. Arena A seats 40,000 people. Arena B seats 50,000. Which arena should host the game? Explain how you would decide.

_____

_____

Lesson 1-1

digit

2,340,581

2, 3, 4, 0, 5, 8, 1

Lesson 1-2

expanded form

$105,073 = 100,000 + 5,000 + 70 + 3$

Lesson 1-3

is equal to (=)

$1,500,000 = 1,500,000$

Lesson 1-3

is greater than (>)

$1,900,000 > 1,700,000$

Lesson 1-3

is less than (<)

$1,200,000 < 1,600,000$

Lesson 1-3

number line

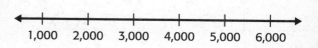

1,000 2,000 3,000 4,000 5,000 6,000

Lesson 1-2

period

Millions Period			Thousands Period			Ones Period		
Millions			Thousands			Ones		
hundreds	tens	ones	hundreds	tens	ones	hundreds	tens	ones
6	5	0,	0	8	4,	9	7	0

Lesson 1-1

place value

Millions			Thousands			Ones		
hundreds	tens	ones	hundreds	tens	ones	hundreds	tens	ones
6	5	0,	0	8	4,	9	7	0

Ideas for Use

- Develop categories for the words. Sort them by category. Ask another student to guess each category.

- Write a tally mark on each card every time you read or write the word. Try to use at least 10 tally marks for each word card.

The representation of a number as a *sum* that shows the value of each digit.

Write the suffix in *expanded*. Use a dictionary if you need help.

_____

Any symbol used to write a whole number.

Write a number in which the digit 6 is in the ones place and the digit 0 is in the ten thousands place.

_____

An inequality relationship showing the number on the left is greater than the number on the right.

List three other math words you can use to compare that end in *-er*.

_____

_____

Having the same value as.

The suffix *-ity* means "state or condition." What does the word *equality* mean?

_____

_____

A line with numbers on it in order at regular intervals.

Write a word problem to compare numbers.

_____

_____

An inequality relationship showing the number on the left is less than the number on the right.

Write a number sentence about two sets of items in the room. Use the symbol < in your sentence.

_____

_____

The value given to a digit by its place in a number.

Write a five-digit number. Then write the place value for each digit.

_____

_____

The name given to each group of three digits in a place-value chart.

What are the names of the three periods on this card?

_____

_____

MY Vocabulary Cards

Mathematical
PRACTICE

standard form

$3,000 + 400 + 90 + 1 = \underbrace{3,491}_{\text{standard form}}$

word form

$16,499 =$ sixteen thousand,
four hundred ninety-nine

Ideas for Use

- Draw or write examples for each card. Be sure your examples are different from what is shown on each card.

- Write the name of each lesson on the front of each blank card. Write a few study tips for each lesson on the back of each card.

The form of a number that uses written words.

Write a five-digit number using numerals. Then rewrite it in word form.

_____

_____

The usual way of writing a number that shows only its *digits*, no words.

One meaning of *standard* is "normal or usual". How can knowing this help you remember the definition for *standard form*?

_____

_____

MY Foldable

FOLDABLES® Follow the steps on the back to make your Foldable.

✂

9	8	7	ones × 1
6	5	4	tens × 10
3	2	1	hundreds × 10 × 10

thousands × 10 × 10 × 10

millions × 10 × 10 × 10 × 10 × 10 × 10

ten thousands × 10 × 10 × 10 × 10

hundred thousands × 10 × 10 × 10 × 10 × 10

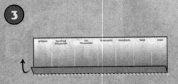

Place Value

Lesson 1

ESSENTIAL QUESTION
How does place value
help represent the value
of numbers?

A **digit** is any symbol used to write a whole
number. The value given to a digit by its place in
a number is called **place value**. A place-value
chart shows the value of the digits in a number.

Math in My World

Tools Watch Tutor

Rhode Island

Example 1

The average pencil can draw a line that is
almost 184,800 feet long. What is the value
of the highlighted digit?

The place-value chart shows 184,800.

Thousands Period			Ones Period		
hundreds	tens	ones	hundreds	tens	ones
1	8	4	8	0	0

100,000
1 × 100,000

4,000
4 × 1,000

0
0 × 10

0
0 × 1

80,000
8 × 10,000

800
8 × 100

The highlighted digit, 8, is in the <u>ten thousands</u> place.

So, its value is <u>80,000</u> .

A digit in each place represents ten times what it would represent
in the place to its right.
When 8 is in the ten thousands place it has a value of 80,000.
If 8 were in the hundred thousands place, it would have a value of
10 × 80,000, or 800,000.

Online content at connectED.mcgraw-hill.com

Example 2

There are 2,419,200 seconds in four weeks. How does the value of the digit in the hundreds place change if this digit were to move to each of the four places to its left?

1,000,000 seconds to go

Millions Period			Thousands Period			Ones Period		
hundreds	tens	ones	hundreds	tens	ones	hundreds	tens	ones
		2	4	1	9	2	0	0

The digit in the hundreds place is ___2___.

It has a value of ___200___.

If this digit were to move to the thousands place, it would have a value of ___2,000___.

If this digit were to move to the ten thousands place, it would have a value of ___20,000___.

If this digit were to move to the hundred thousands place, it would have a value of ___200,000___.

If this digit were to move to the millions place, it would have a value of ___2,000,000___.

The digit in each place has a value that is ten times as great as it has in the place to its right.

Guided Practice Check ✓

Circle the correct place of the highlighted digit and write its value.

		Place		Value
1.	62,574	(ones)	tens	___0___
2.	53,456	(ten thousands)	thousands	___50,000___
3.	59,833	tens	(hundreds)	___
4.	174,305	ten thousands	thousands	___

Talk MATH

How does the value of a digit in the thousands place compare to its value if the same digit was in the hundreds place?

Independent Practice

Circle the place of the highlighted digit and write its value.

		Place		**Value**
5. 593,802	hundreds		tens	0
6. 4,826,193	ten thousands		hundred thousands	800,000
7. 7,830,259	hundred thousands		millions	7,000,000

Use the place-value chart for Exercises 8–16.

Thousands Period			Ones Period		
hundreds	tens	ones	hundreds	tens	ones
4	6	2	3	7	1

8. The 6 is in the ___ten thousands___ place.

9. The ___2___ is in the thousands place.

10. The 7 has a value of 7 × ___10___.

11. The 6 has a value of 6 × ___10,000___.

12. The ___4___ has a value of ___4___ × 100,000.

13. The ___3___ is in the hundreds place.

14. The 1 is in the ___ones___ place.

15. The digit in the hundred thousands place has 10 times the value

it would have if it was in the ___ten thousands___ place.

16. The digit in the thousands place has _____ times the
value it would have if it was in the hundreds place.

Problem Solving

17. An African elephant can weigh up to 14,432 pounds. What is the value of the highlighted digit?

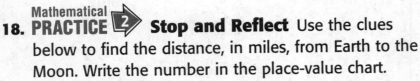

18. Mathematical **PRACTICE** ➋ **Stop and Reflect** Use the clues below to find the distance, in miles, from Earth to the Moon. Write the number in the place-value chart.

- The greatest place value is hundred thousands.
- The digit in the tens place is 5.
- The remaining digits are 2, 3, 8, and 7.
- One of the digits has a value of 30,000.
- One of the digits has a value of 800. The value of the digit in the thousands place is 10 times greater than this.
- There are 2 more ones than tens.

Thousands Period			Ones Period		
hundreds	tens	ones	hundreds	tens	ones

What is the distance from Earth to the Moon?

_____

My Work!

HOT Problems

19. Mathematical **PRACTICE** ➍ **Model Math** Write a six-digit number that has a 9 in the hundreds place and a 6 in the hundred thousands place.

_____

20. ❓ **Building on the Essential Question** How does moving the place of a digit change its value?

_____

_____

_____

MY Homework

Homework Helper

Need help? ↗ connectED.mcgraw-hill.com

Write the place and value of the highlighted digit in 8,304,421.

Use a place-value chart.

Millions Period			Thousands Period			Ones Period		
hundreds	tens	ones	hundreds	tens	ones	hundreds	tens	ones
		8	3	0	4	4	2	1

The **3** is in the hundred thousands place.

The value of the **3** is 3 × 100,000, or 300,000.

Practice

Circle the place of the highlighted digit and write its value.

		Place		Value
1. 62,468	thousands		ten thousands	_____
2. 934,218	thousands		ten thousands	_____
3. 438,112	ten thousands		hundred thousands	_____
4. 285,012	tens		thousands	_____
5. 2,905,146	hundred thousands		millions	_____
6. 6,034,215	ten thousands		millions	_____

Problem Solving

Mathematical PRACTICE **2** **Use Number Sense**

Use the place-value chart for Exercises 7–13.

Millions Period			Thousands Period			Ones Period		
hundreds	tens	ones	hundreds	tens	ones	hundreds	tens	ones

7. Write 1 in the hundreds place.

8. Write 8 in the tens place.

9. Write 4 in the ones place.

10. Write 3 in the thousands place.

11. Write 7 in the millions place.

12. Write 5 in the ten thousands place.

13. Write 2 in the hundred thousands place.

Vocabulary Check

Match each definition to the correct vocabulary term.

14. The value given to a digit by its position in a number.

    • digits

15. Symbols used to write whole numbers.

    • place value

Test Practice

16. A digit is in the hundreds place. The digit is moved so that its value is ten times greater. To which place did the digit move?

    Ⓐ hundred thousands

    Ⓑ ten thousands

    Ⓒ thousands

    Ⓓ tens

Read and Write Multi-Digit Numbers

Lesson 2

ESSENTIAL QUESTION
How does place value help represent the value of numbers?

Place-value charts show the value of each digit. A group of three digits is called a **period**. Commas separate the periods. At each comma, say the name of the period.

Math in My World

Example 1

Scientists found that a gooney bird once traveled 24,983 miles in just 90 days. Use a place-value chart to read the number of miles that the gooney bird traveled.

The place-value chart shows 24,983.

Thousands Period			Ones Period		
hundreds	tens	ones	hundreds	tens	ones
	2	4	9	8	3

The comma is after the thousands period. Say *thousand* when reading the number.

Say: *twenty-four thousand, nine hundred eighty-three*

So, the gooney bird traveled *twenty-four thousand, nine hundred eighty-three* miles.

There are different ways to write numbers. **Standard form**, or number names, only uses digits to show the number. **Expanded form** shows the number as a sum of the values of each digit. **Word form** shows the number only using words.

Online content at ⤸ **connectED.mcgraw-hill.com**

Example 2

The population of Botswana is about 1,882,000. Write this number in standard form, expanded form, and word form.

The place-value chart shows the place of each digit.

Millions Period			Thousands Period			Ones Period		
hundreds	tens	ones	hundreds	tens	ones	hundreds	tens	ones
		1	8	8	2	0	0	0

Write the number in standard form and expanded form.

☐ , 8 ☐ ☐ , 0 ☐ ☐ ←  standard form

1,000,000 + _____ + 80,000 + _____ ←  expanded form

Write the number in word form.

one million, _____ *hundred eighty-* _____ *thousand*

Guided Practice

Write each number in standard form.

1. *three hundred forty-nine thousand, twenty-five* _____

2. 400,000 + 90,000 + 2,000 + 800 + 10 + 4 _____

Write each number in expanded form and word form.

3. 492,032

_____

_____

4. 3,028,002

_____

_____

Talk MATH

What is the value of the 6 in 345,629?

Independent Practice

Write each number in standard form.

5. *twenty-five thousand, four hundred eight* _____

6. *forty thousand, eight hundred eleven* _____

7. *seven hundred sixty-one thousand, three hundred fifty-six* _____

8. *five million, seven hundred sixty-two thousand, one hundred eleven*

_____

9. 600,000 + 80,000 + 4 _____

10. 20,000 + 900 + 70 + 6 _____

11. 9,000,000 + 200,000 + 1,000 + 500 + 2 _____

Write each number in expanded form and word form.

12. 485,830

Expanded Form:

_____

Word Form:

_____

_____

13. 3,029,251

Expanded Form:

_____

Word Form:

_____

Problem Solving

14. **Mathematical PRACTICE 1** **Plan Your Solution** A zoo's newborn African elephant weighed 232 pounds. After one year, the elephant gained 1,000 pounds. Write the elephant's new weight in expanded form and word form.

Expanded form: _____

Word Form: _____

_____

15. The population of Norway is about *four million, seven hundred two thousand*. Write this number in standard form.

_____

16. The population of the Dominican Republic is about 9,366,000. Write this number in word form.

_____

_____

My Work!

HOT Problems

17. **Mathematical PRACTICE 3** **Find the Error** Sonia wrote the expanded form of 2,408,615 below.

$$2,000,000 + 400,000 + 80,000 + 600 + 10 + 5$$

Find and correct her mistake.

_____

_____

18. **Building on the Essential Question** Why is expanded form important? Explain.

_____

MY Homework

Homework Helper

Need help? connectED.mcgraw-hill.com

Write 1,000,000 + 300,000 + 60,000 + 300 + 10 + 5 in standard form. Then read the number aloud.

standard form: 1,360,315

Remember: Commas separate the periods. Say the name of the period at each comma.

Millions Period			Thousands Period			Ones Period		
hundreds	tens	ones	hundreds	tens	ones	hundreds	tens	ones
		1	3	6	0	3	1	5

Say: *one million, three hundred sixty thousand, three hundred fifteen*

Write 756,491 in expanded form and word form.

expanded form: 700,000 + 50,000 + 6,000 + 400 + 90 + 1

word form: *seven hundred fifty-six thousand, four hundred ninety-one*

Practice

1. Write *one million, one hundred forty-five thousand, two hundred thirty-seven* in standard form.

_____

2. Write 87,192 in word form and expanded form.

_____

_____

Problem Solving

Complete the expanded form.

3. Mathematical PRACTICE Check for Reasonableness

91,765 = 90,000 + _____ + 700 + _____ + 5

4. 798,054 = 700,000 + _____ + _____ + 50 + 4

5. 5,925,020 = 5,000,000 + _____ + 20,000 + _____ + 20

6. 2,802,136 = _____ + 800,000 + _____ + 100 + 30 + _____

Vocabulary Check

Read each definition. Choose the correct word(s) to fill in the spaces.

expanded form   period

standard form   word form

7. the way of writing a number using words

___ ___ ___ ___   ___ ___ ___ ___

8. the usual way of writing a number, using digits

___ ___ ___ ___ ___ ___ ___ ___

9. the way of writing a number as the sum of the value of each digit

___ ___ ___ ___ ___ ___ ___ ___ ___ ___

10. each group of three digits on a place-value chart

___ ___ ___ ___ ___ ___

Test Practice

11. Which is the correct expanded form for 45,098?

Ⓐ 45,000 + 98

Ⓑ 4,000 + 5,000 + 9 + 8

Ⓒ 40,000 + 500 + 90 + 8

Ⓓ 40,000 + 5,000 + 90 + 8

Compare Numbers

Lesson 3
ESSENTIAL QUESTION
How does place value help represent the value of numbers?

A **number line** is a line with numbers on it in order at regular intervals. You can use a number line to compare numbers. Use these symbols to show how numbers relate to each other.

| **is greater than (>)** | **is less than (<)** | **is equal to (=)** |

 Math in My World Tools Watch Tutor

Example 1

On average, a first-year police officer earns $41,793. A first-year firefighter earns $41,294. Which occupation pays more the first year?

Label each dot with the correct salary.

41,793 is to the

right of 41,294.

41,793 is _greate_ than 41,294.

41,793 (>) 41,294 ← Write the > symbol in the circle.

So, _police_ earn more the first year than _fire_ .

41,000 41,200 41,400 41,600 41,800 42,000

◄ Numbers get smaller.

Numbers get larger. ►

Example 2

In a recent year, the population of Vermont was 621,760. The population of North Dakota was 646,844. Compare these two populations. Use <, >, or =.

Use a place-value chart.

1 Write the numbers on the place-value chart.

2 Compare the digits of the greatest place value. If they are the same, move to the next digit to the right until you find digits that are different.

Thousands Period			Ones Period		
hundreds	tens	ones	hundreds	tens	ones

same   different

2 < 4

So, 621,760 < 646,844.

Guided Practice

1. Use the number line to compare. Use <, >, or =.

32,053 < 35,251

30,000 31,000 32,000 33,000 34,000 35,000 36,000

Compare. Use <, >, or =.

2. 25,409 < 26,409

3. 13,636 = 13,636

4. 72,451 > 76,321

5. 201,953 = 201,953

6. 442,089 > 442,078

7. 224,747 > 224,774

Talk MATH

If two numbers have all of the same digits in the same places, can one of them be greater than the other? Explain.

Name ....

Independent Practice

For Exercises 8–10, use the number lines to compare.
Use <, >, or =.

8. 45,526 (>) 48,873

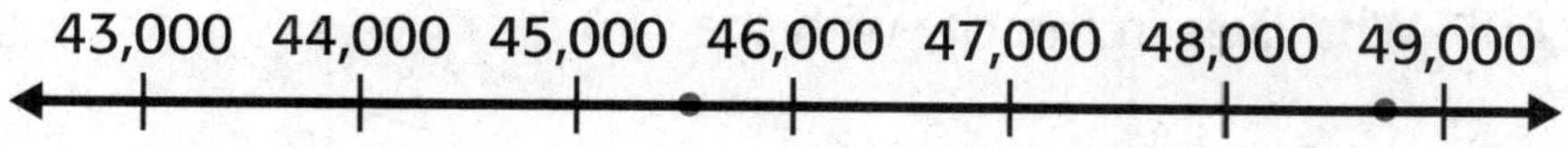
43,000
44,000
45,000
46,000
47,000
48,000
49,000

9. 31,748 (<) 31,521

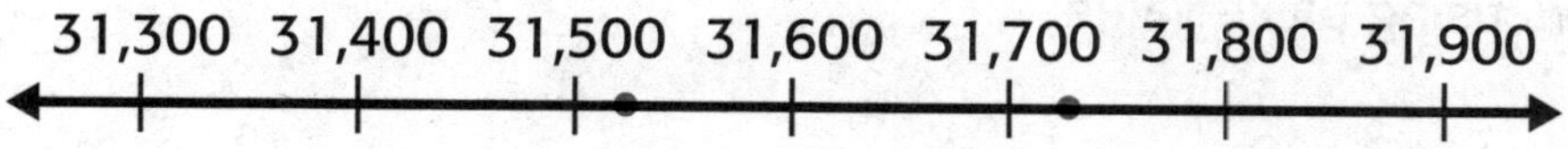
31,300
31,400
31,500
31,600
31,700
31,800
31,900

10. 126,532 (<) 129,321

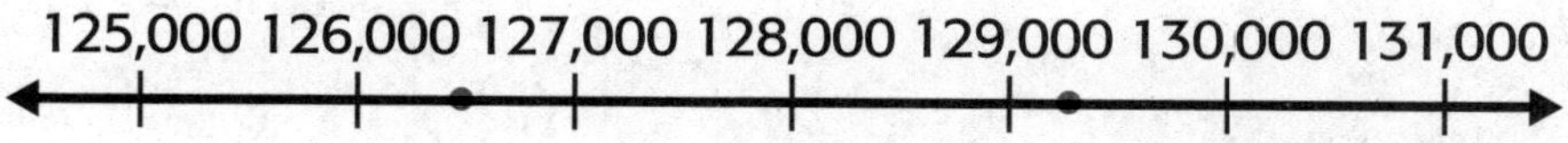
125,000
126,000
127,000
128,000
129,000
130,000
131,000

Compare. Use <, >, or =.

11. 3,030 (=) 3,030 **12.** 76,101 (>) 77,000 **13.** 12,683 (<) 12,638

14. 229,214 (>) 300,142 **15.** 701,000 (=) 701,000 **16.** 342,646 (<) 34,646

17. 398,421 (<) 389,421 **18.** 605,310 (=) 605,310 **19.** 840,515 (<) 845,015

20. 655,543 (<) 556,543 **21.** 720,301 (>) 720,031 **22.** 333,452 (=) 333,452

Problem Solving

23. Jun collects stamps and baseball cards. He has 1,834 stamps and 1,286 baseball cards. Does he have more stamps or more cards? Explain.

he has more staps

24. The population of Cindy's city is 242,506. The population of Mark's city is 242,605. Who lives in the city with the greater population?

cindy city

25. Mathematical PRACTICE 6 **Explain to a Friend** Explain how to compare numbers using place value.

ten's

HOT Problems

Mathematical PRACTICE 2 **Understand Symbols** For Exercises 26–28, fill in each blank to make the number sentence true.

26. 253,052 < _____

27. 95,925 > _____

28. 205,053 < _____

29. ❓ **Building on the Essential Question** How can I show how numbers are related to each other?

_____

_____

_____

MY Homework

Homework Helper

Need help? connectED.mcgraw-hill.com

Compare 54,515 and 54,233. Use >, <, or =.

Use a number line.

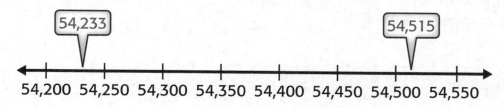

54,233   54,515

54,200 54,250 54,300 54,350 54,400 54,450 54,500 54,550

54,515 is to the right of 54,233 on the number line.

So, 54,515 > 54,233.

Practice

For Exercises 1–2, use the number lines to compare. Use <, >, or =.

1. 67,113 ◯ 62,523

62,000 63,000 64,000 65,000 66,000 67,000 68,000

2. 42,254 ◯ 42,533

42,000 42,100 42,200 42,300 42,400 42,500 42,600

Compare. Use <, >, or =.

3. $751,012 ◯ $715,012   **4.** 4,350 ◯ 5,430   **5.** 8,080 ◯ 8,880

6. 322,650 ◯ 332,650   **7.** 673 ◯ 376   **8.** $918,050 ◯ $819,050

9. 121,571 ◯ 211,571   **10.** 17,888 ◯ 17,780   **11.** 72,770 ◯ 72,770

Problem Solving

12. Mathematical **PRACTICE** ➌ **Draw a Conclusion** Gigi has $1,698 in her savings account. Robert has $1,898 in his savings account. Toby has $100 less than Robert in his savings account. Who has the least amount of money?

_____

13. There were 544,692 tickets sold for the rock concert. There were 455,692 tickets sold for the country music concert. Which concert sold a greater number of tickets?

_____

Vocabulary Check

14. Choose the correct word(s) to complete each sentence.

is equal to (=)	is greater than (>)
is less than (<)	number line

To compare numbers, you can use a _____. A number that is to the right on a number line _____ a number to its left. A number on the left _____ a number to its right.

You can look at place values to compare numbers. If a number has a digit in the thousands place that _____ the thousands digit in another number, then look to the hundreds place.

Test Practice

15. Which number sentence is *not* true?

Ⓐ 243,053 < 242,553

Ⓑ 194,832 > 193,832

Ⓒ 553,025 = 553,025

Ⓓ 295,925 < 295,952

Order Numbers

Lesson 4

ESSENTIAL QUESTION
How does place value help represent the value of numbers?

You can use a place-value chart to put numbers in order.

 Math in My World Tutor

Example 1

Compare the populations of the three cities and order them from *greatest* to *least*.

Lowell pop. 103,299
Cambridge pop. 101,365
Boston pop. 590,763

1. Write the populations on the place-value chart.

2. Start with the greatest place-value position. Compare.

   _____ > _____

3. Compare the digits in the next place.

   _____ = _____

4. Continue to compare until the digits are different.

   _____ > _____

   _____ > _____ > _____

So, from *greatest* population to *least*, the order of the cities is

_____ , _____ , and _____ .

Thousands			Ones		
hundreds	tens	ones	hundreds	tens	ones
			,		
			,		
			,		

Example 2

Order the numbers on the cards at the right from *least* to *greatest*.

245,032 254,002 245,023

1 Line up the numbers by the ones place.

☐☐☐,☐☐☐
☐☐☐,☐☐☐
☐☐☐,☐☐☐

2 Start with the greatest place-value position. Compare.

Each number has a ＿＿＿＿ in the hundred thousands place. So, compare the digits in the ten thousands place. The greatest number is ＿＿＿＿.

The two numbers that are left both have a ＿＿＿＿ in the thousands place and a ＿＿＿＿ in the hundreds place. The second greatest number is ＿＿＿＿.

So, in order from *least* to *greatest*, the numbers are

＿＿＿＿＿＿＿＿＿＿＿＿＿＿＿.

Talk MATH

When ordering numbers, what do you do when the digits in the same place have the same value?

Guided Practice ✓

1. Place the numbers in the place-value chart in order from *greatest* to *least*.

   52,482
   50,023
   56,028
   63,340

Thousands			Ones		
hundreds	tens	ones	hundreds	tens	ones
greatest →					
least →					

Name

Independent Practice

Place the numbers in the place-value chart in order from *greatest* to *least*.

2. 12,378
12,783
12,873

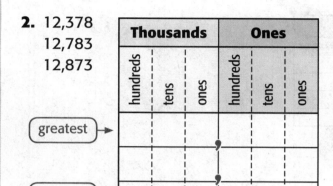

greatest →
least →

3. 258,103
248,034
285,091
248,934

greatest →
least →

4. 138,032
138,023
139,006
183,467

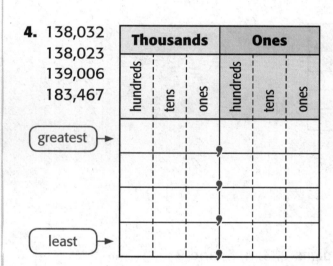

greatest →
least →

5. 652,264
625,264
652,462
625,642

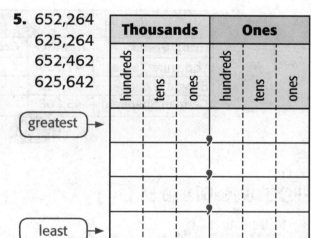

greatest →
least →

Order the numbers from *least* to *greatest*.

6. 402,052; 425,674; 414,035

7. 643,947; 643,537; 642,066

8. 113,636; 372,257; 337,633

9. 563,426; 564,376; 653,363

Problem Solving

10. **Mathematical PRACTICE 2** **Stop and Reflect** Order the states from *least* (1) to *greatest* (4) total area.

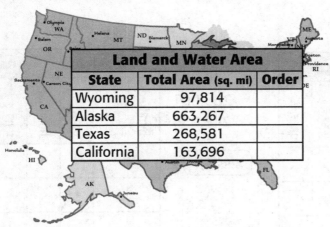

Land and Water Area		
State	**Total Area (sq. mi)**	**Order**
Wyoming	97,814	
Alaska	663,267	
Texas	268,581	
California	163,696	

11. Order the dog breeds from *least* popular (1) to *most* popular (3).

Dog Breeds		
Dog Breed	**Number**	**Order**
Yorkshire Terrier	47,238	
Beagle	42,592	
German Shepherd	45,868	

My Work!

HOT Problems

12. **Mathematical PRACTICE 1** **Keep Trying** Use the digits 2, 3, 4, 5, and 9 to create four 5-digit numbers. Use each digit exactly once in each number. Order them from *least* to *greatest*.

_____

_____

13. **Building on the Essential Question** When do I compare real-world numbers?

_____

_____

Name ....

MY Homework

Lesson 4
Order Numbers

Homework Helper

Need help? connectED.mcgraw-hill.com

Order the numbers from *greatest* to *least*:
17,601; 20,007; 17,610

Compare the ten thousands.

17,601
20,007 ◄——— most ten thousands
17,610

**Both thousands and hundreds are the same,
so compare the tens.**

17,601
17,610 ◄——— more tens

From *greatest* to *least*, the numbers are 20,007; 17,610; and 17,601.

Practice

Order the numbers from *greatest* to *least*.

1. 59,909; 95,509; 59,919

2. 2,993; 9,239; 2,393

3. 112,443; 114,324; 112,344

4. 642,063; 642,036; 642,306

Order the numbers from *least* to *greatest*.

5. 225,625; 335,432; 325,745

6. 357,925; 329,053; 356,035

Problem Solving

7. The United States' soccer team has 572,112 fans. Great Britain's team has 612,006 fans. Brazil's team has 901,808 fans. Write the countries in order from the *greatest* to *least* number of soccer fans.

_____

_____

8. There are 943,025 sports tickets available. There are 832,502 movie tickets available. There are 415,935 theater tickets available. List the number of tickets in order from *least* to *greatest*.

_____

_____

9. **Mathematical PRACTICE 7 Identify Structure** Write four numbers that each have six digits. Order the numbers from *least* to *greatest*.

_____

_____

Test Practice

10. The table shows the populations of the cities where Alex and Brent live. Marcia lives in a city that has more people than Alex's city and fewer people than Brent's. Which could be the number of people who live in Marcia's city?

Name	Population of their city
Alex	404,048
Brent	412,888

   Ⓐ 413,066 people   Ⓒ 404,132 people

   Ⓑ 412,901 people   Ⓓ 403,997 people

Check My Progress

Vocabulary Check

1. Each card shows the definition or an example of a vocabulary word. Write each word from the word bank on the card with the matching definition or example.

digit **expanded form** **is equal to (=)** **is greater than (>)**

is less than (<) **number line** **period** **place value**

standard form **word form**

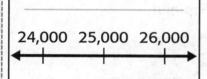
24,000 25,000 26,000

23,042 ◯ 23,000

Example: 83,104

---

the value given to a digit by its position in a number

---

any symbol used to write a whole number

---

the name given to each group of three digits on a place-value chart

---

Example:
80,000 + 3,000 + 100 + 4

34,842 ◯ 43,842

Example:
eighty-three thousand, one hundred four

44,204 ◯ 44,204

Concept Check ✓

Write the place of the highlighted digit. Then write its value.

2. 34,025

3. 52,276

_____

_____

_____

_____

Problem Solving

4. A car costs *thirty-six thousand, five hundred forty-seven* dollars. Write the cost in standard form.

_____

5. A farm sold 429,842 apples and 53,744 pears. Did the farm sell more apples or pears?

_____

6. The table shows how much money was made at Laser Flip and Grind's Skateboard Store.

Laser Flip and Grind's Skateboard Store Sales	
Item	**Sales**
Skateboards	$132,439
Helmets	$103,322
Ramps	$201,385

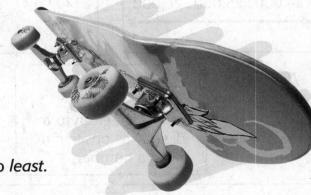

Write the sales in order from *greatest* to *least*.

_____

Test Practice

7. Which set of numbers is written from *least* to *greatest*?

Ⓐ 351,935; 351,914; 215,634

Ⓑ 351,914; 215,634; 351,935

Ⓒ 215,634; 351,935; 351,914

Ⓓ 215,634; 351,914; 351,935

Number and Operations in Base Ten
4.NBT.3

CCSS

Use Place Value to Round

Lesson 5

ESSENTIAL QUESTION
How does place value help represent the value of numbers?

When you estimate, you find an answer that is close to the exact answer. One way to estimate is to round by changing the value of a number so it is easier to work with.

 Math in My World Tools Watch Tutor

Example 1

The largest extreme sports competition, called X-Games, is so popular that one year, 268,390 people attended. What is 268,390 rounded to the nearest ten thousand?

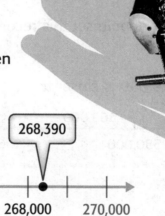

Look at the number line. 268,390 is between 260,000 and 270,000.

268,390

260,000   262,000   264,000   266,000   268,000   270,000

Since 268,390 is closer to _____ than

_____ , 268,390 rounds to _____ .

Example 2

A record was set when 569,069 people jumped up and down for one minute. About how many people set this record? Round 569,069 to the nearest hundred thousand.

 Circle the digit in the place to be rounded.

569,069

2 Underline the digit to the right of the place being rounded.

3 If the underlined digit is 4 or less, do not change the circled digit. If the underlined digit is 5 or greater, add 1 to the circled digit.

4 Replace all digits after the circled digit with zeros.

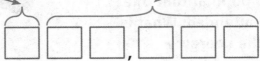

So, about _____ people set this record.

Check

Use a number line. 569,069 is closer to _____ than 500,000.

569,069

$\longleftarrow$ |————————|————————|————————| $\longrightarrow$
500,000   550,000   600,000

Guided Practice

Round to the given place-value position.

1. 2,221; thousands _____

2. 78,214; ten thousands _____

3. 581,203; hundred thousands _____

Talk MATH

What is the least number that you can round to the thousands place to get 8,000? Explain.

Independent Practice

Round to the given place-value position.

4. 500,580; thousands

5. 290,152; hundred thousands

6. 218,457; hundred thousands

7. 37,890; hundreds

8. 95,010; thousands

9. 845,636; ten thousands

10. 336,001; hundred thousands

11. 709,385; hundred thousands

Tell the place-value position to which each number was rounded.

12. 456,750 ⟶ 460,000

13. 38,124 ⟶ 38,120

14. 18,334 ⟶ 18,000

15. 455,670 ⟶ 455,700

16. 980,065 ⟶ 980,070

17. 162,245 ⟶ 200,000

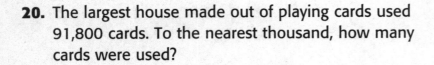

Problem Solving

A natural gas and fuel-friendly car was used to set a world record.

I'm GREEN!

World Record
23,697 miles or
38,137 kilometers

18. What is the distance in miles traveled to the nearest ten thousand?

_____

19. Round the distance traveled in kilometers to the nearest thousand.

_____

My Work!

20. The largest house made out of playing cards used 91,800 cards. To the nearest thousand, how many cards were used?

_____

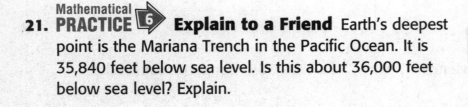

21. Mathematical **PRACTICE** 6 **Explain to a Friend** Earth's deepest point is the Mariana Trench in the Pacific Ocean. It is 35,840 feet below sea level. Is this about 36,000 feet below sea level? Explain.

_____

_____

HOT Problems

22. Mathematical **PRACTICE** 3 **Find the Error** Andrew rounded the number 672,726 to the nearest hundred thousand. He wrote 672,000. Find and correct his mistake.

_____

_____

23. **Building on the Essential Question** When is estimation an effective way to determine an answer?

_____

_____

MY Homework

Homework Helper

Need help? connectED.mcgraw-hill.com

Round 65,839 to the nearest hundred.

Circle the digit to be rounded.   65,⑧39

The digit to the right is 4 or less, so the 8 does not change. All digits after the 8 are replaced with zeros.

65,839 rounded to the nearest hundred is 65,800.

Round 65,839 to the nearest ten thousand.

Circle the digit to be rounded.   ⑥5,839

The digit to the right is 5 or more, so 1 is added to the circled digit. The digits after the circled digit are replaced with zeros.

65,839 rounded to the nearest ten thousand is 70,000.

Practice

Round each number to the given place-value position.

1. 64,569; thousands

_____

2. 155,016; thousands

_____

3. 73,569; ten thousands

_____

4. 708,569; ten thousands

_____

5. 91,284; hundred thousands

_____

6. 265,409; hundred thousands

_____

Problem Solving

7. Luis and his family flew 51,487 miles last summer while on vacation. Rounded to the nearest thousand, how many miles did they fly?

_____

8. Miles bought a car that cost $23,556. To the nearest ten thousand, how much did the car cost?

_____

9. Explain how you would round the numbers 33 and 89 to estimate their sum.

_____

_____

_____

Use the data from the table for Exercises 10–12.

10. **Mathematical PRACTICE 5** **Use Math Tools** Which ocean has an average depth of about 12,000 feet, to the nearest thousand?

Depths of Oceans	
Ocean	Average Depth (ft)
Pacific	12,925
Atlantic	11,730
Indian	12,598

_____

11. What is the depth of the Pacific Ocean rounded to the nearest ten thousand?

_____

12. What is the depth of the Indian Ocean rounded to the nearest thousand?

_____

Test Practice

13. What is 104,229 rounded to the nearest ten thousand?

Ⓐ 90,000   Ⓒ 104,000

Ⓑ 100,000   Ⓓ 110,000

Problem-Solving Investigation

STRATEGY: Use the Four-Step Plan

Lesson 6

ESSENTIAL QUESTION
How does place value help represent the value of numbers?

Learn the Strategy

Ben, Andy, and Kelly each live in a different city. The populations of the cities are 372,952; 225,395; and 373,926. Use the clues to find the population of the city where Ben lives.

Clues
- Andy's city has the least population.
- When rounded to the nearest thousand, the population of Kelly's city is 374,000.

1 Understand

What facts do you know?

Ben, Andy, and Kelly each live in a different city.

The populations for each city are: _____ ; _____ ; and _____ .

What do you need to find?

the population of _____ city

2 Plan

I can order and round the populations.

3 Solve

Order the populations from *least* to *greatest*. 225,395; 372,952; 373,926

_____ lives in the city with the least population.

Round the remaining populations to the nearest thousand.

372,952 rounds to _____ . 373,926 rounds to _____ .

Kelly lives in the city with the population that rounds to 374,000.

So, Ben must live in the city that has a population of _____ .

4 Check

Does your answer make sense? Explain.

Practice the Strategy

It is estimated that a movie made more than $7,000,000 but less than $8,000,000. There is a 5 in the thousands place, a 7 in the ten thousands place, and a 6 in the hundred thousands place. The ones, tens, and hundreds places have zeros in them because the total is an estimate. What is the estimated amount of money that the movie made?

 Understand

What facts do you know?

_____

_____

What do you need to find?

_____

2 Plan

3 Solve

4 Check

Does your answer make sense? Explain.

_____

_____

Apply the Strategy

What a deal!

Solve each problem by using the four-step plan.

1. **Mathematical** **PRACTICE** 5 **Use Math Tools** Mr. Kramer is buying a car. The list of prices is shown in the table.

   Mr. Kramer wants to buy the least expensive car.

   Which car should he buy?

Prices of Cars	
Cars	Price
Car A	$83,532
Car B	$24,375
Car C	$24,053
Car D	$73,295

My Work!

2. A restaurant made more than $80,000 but less than $90,000 last month. There is a 6 in the ones place, a 3 in the thousands place, a 7 in the hundreds place, and a 1 in the tens place. How much money did the restaurant make last month?

3. Amy, Lisa, Angie, and Doug live in different states. The populations of those states are 885,122; 5,024,748; 4,492,076; and 2,951,996. Lisa lives in the state with the greatest population. Doug lives in the state with a 2 in the thousands place of the population. Amy lives in the state with the least population. What is the population of Angie's state?

4. The New Meadowlands Stadium in New Jersey seats a large number of fans. There are zeros in the tens and ones places, a 2 in the thousands place, an 8 in the ten thousands place, and a 5 in the hundreds place. How many people does the stadium seat?

Review the Strategies

Use any strategy to solve each problem.
- Make a table.
- Choose an operation.
- Act it out.
- Draw a picture.

5. The table at the right shows which types of canned foods were collected during a food drive.

What was the most popular type of canned food collected?

Canned Foods	Number Collected
tomatoes	59,294
beans	159,002
corn	45,925
soup	903,690

My Work!

_____

6. The population of a city has 6 digits. There is a 3 in the tens place, a 5 in the hundred thousands place, a 6 in the ones place, and a 9 in the rest of the places. What is the population of the city?

_____

7. A warehouse stores cans of paint. There is a 3 in the hundreds place, a 7 in the thousands place, a 5 in the ten thousands place, and an 8 in the rest of the places. This number has 5 digits. How many cans of paint are in the warehouse?

_____

8. Mathematical **PRACTICE** 7 **Identify Structure** A car's mileage is a five-digit number. There is a 3 in the ten thousands place, the ones place, and the tens place. There is a 9 in the hundreds place and the thousands place. What is the car's mileage?

_____

Name ....

MY Homework

Homework Helper eHelp

Need help? connectED.mcgraw-hill.com

A six-digit number has a 2 in the thousands place, a 5 in the tens place, a 3 in the hundred thousands place, and zeros in each of the remaining places. What is the number?

Use the four-step plan to solve this problem.

1 Understand

I know that there is a number with six digits. It has a 2 in the thousands place, a 5 in the tens place, a 3 in the hundred thousands place, and zeros in each of the remaining places. I need to find the number.

2 Plan

I will use a place-value chart to help me organize the digits.

3 Solve

Thousands Period			Ones Period		
hundreds	tens	ones	hundreds	tens	ones
3	0	2,	0	5	0

So, the number is 302,050.

4 Check

I can check my work by reading the clues again to make sure that the digits are all in the correct places.

Problem Solving

1. A five-digit number has a 3 in the hundreds place, a 7 in the greatest place-value position, a 9 in the ones place, an 8 in the thousands place, and a 6 in the tens place. What is the number? Use the four-step plan.

Solve each problem by using the four-step plan.

2. Use the digits 1–7 to create a seven-digit number that can be rounded to 6,300,000.

_____

3. A seven-digit number has a 0 in the ones place, a 6 in the ten thousands place, an 8 in the millions place, and fives in each of the remaining places. What is the number?

_____

4. Tara rolled the numbers shown. What is the greatest number she can make using each digit once?

_____

5. Betsy, Carl, and Dave each live in different cities. The populations of the cities are 194,032; 23,853; and 192,034. Betsy lives in the city with the least population. Carl does not live in the city with the greatest population. What is the population of Dave's city?

_____

Mathematical
6. PRACTICE 6 **Explain to a Friend** Explain how the value of the 7 in 327,902 will change if you move it to the tens place.

_____

_____

Vocabulary Check

Use the words in the word bank to complete each sentence.

digits	expanded form	is equal to (=)
is greater than (>)	is less than (<)	number line
period	place value	standard form
word form		

1. 83,502 _____ 82,502.

2. You can use a _____ to compare numbers.

3. There are five _____ in the number 35,024.

4. 392,903 _____ 392,903.

5. The _____ of 32,052 is *thirty-two thousand, fifty-two*.

6. The _____ of 853,025 is 800,000 + 50,000 + 3,000 + 20 + 5.

7. _____ is the value given to a digit by its position in a number.

8. The name given to each group of three digits on a place-value chart is called a _____ .

9. The _____ of *fifteen thousand, sixty-two* is 15,062.

10. 473,503 _____ 474,503.

Concept Check ✓

11. Write *two hundred thirty-nine thousand, eight hundred four* in standard form and expanded form.

_____

_____

Compare. Use <, >, or =.

12. 689,000 ◯ 679,000

13. 515,063 ◯ 515,603

14. 739,023 ◯ 739,023

15. 405,032 ◯ 450,002

16. Round 415,203 to the thousands place. _____

Use the place-value chart for Exercises 17–23.

Thousands Period			Ones Period		
hundreds	tens	ones	hundreds	tens	ones
5	3	7	2	8	0

17. The 3 is in the _____ place.

18. The _____ is in the thousands place.

19. The 8 has a value of 8 × _____ .

20. The 3 has a value of 3 × _____ .

21. The _____ has a value of _____ × 100,000.

22. The _____ is in the hundreds place.

23. The digit in each place has a value that is _____ times as great as it has in the place to its _____ .

Order the numbers from *greatest* to *least*.

24. 374,273 _____

374,372 _____

347,732 _____

25. 263,224 _____

623,224 _____

633,222 _____

50 Chapter 1 Place Value

Name ....

 # Problem Solving

26. There were 48,566 people at a Sunday football game. To the nearest thousand, how many people were at the game?

_____

27. The table shows the cost of three houses. Order these prices from least to greatest.

House	Price
House A	$175,359
House B	$169,499
House C	$179,450

_____

28. In a recent year, the population of Hong Kong was about 6,924,000. What is the value of the 9 in this number?

_____

29. A baseball stadium has 24,053 seats. A football stadium has 53,025 seats. Which stadium has a greater number of seats?

_____

Test Practice

30. The population of New Zealand is about 4,184,000. What is the expanded form of this number?

Ⓐ 4,000,000 + 100,000 + 80,000 + 4,000

Ⓑ 4,000,000 + 100,000 + 8,000 + 4,000

Ⓒ 400,000 + 100 + 80 + 4

Ⓓ 4 + 1 + 8 + 4

Reflect

Use what you learned about place value to
complete the graphic organizer.

Write the Example	Real-World Example
Vocabulary	Estimate

ESSENTIAL QUESTION

How does place value
help represent the
value of numbers?

Reflect on the ESSENTIAL QUESTION Write your answer below.

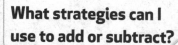

ESSENTIAL QUESTION

What strategies can I use to add or subtract?

Let's Watch the Show!

Watch

Watch a video!

4.NBT.3 Use place value understanding to round multi-digit whole numbers to any place.

4.NBT.4 Fluently add and subtract multi-digit whole numbers using the standard algorithm.

Operations and Algebraic Thinking *This chapter also addresses these standards:*

4.OA.3 Solve multistep word problems posed with whole numbers and having whole-number answers using the four operations, including problems in which remainders must be interpreted. Represent these problems using equations with a letter standing for the unknown quantity. Assess the reasonableness of answers using mental computation and estimation strategies including rounding.

4.OA.5 Generate a number or shape pattern that follows a given rule. Identify apparent features of the pattern that were not explicit in the rule itself.

Standards for
Mathematical
PRACTICE

I'll be able to get this – no problem!

1. Make sense of problems and persevere in solving them.
2. Reason abstractly and quantitatively.
3. Construct viable arguments and critique the reasoning of others.
4. Model with mathematics.
5. Use appropriate tools strategically.
6. Attend to precision.
7. Look for and make use of structure.
8. Look for and express regularity in repeated reasoning.

 = focused on in this chapter

Name

....

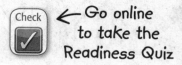

Check ☑ ← Go online to take the Readiness Quiz

Add.

1. 35
    + 56

2. $58
    + $25

3. 94
    + 78

4. $87 + $35 = _____

5. 103 + 57 = _____

6. 233 + 158 = _____

7. Felicia has a collection of 117 marbles. Her sister gives her 25 marbles. How many marbles does Felicia have now?

_____

Subtract.

8. $57
    − $8

9. 71
    − 23

10. 132
     − 74

11. 93 − 15 = _____

12. $62 − $49 = _____

13. 415 − 107 = _____

14. Jasper is reading a 98-page book. He has read 29 pages. How many pages does Jasper have left to read?

_____

Shade the boxes to show the problems you answered correctly.

How Did I Do?

1	2	3	4	5	6	7	8	9	10	11	12	13	14

Online Content at connectED.mcgraw-hill.com

Review Vocabulary

difference	estimate	round	sum	word form

Making Connections

Use the review vocabulary to describe the examples based on the problems in each chart.

Subtraction Problem	Vocabulary Word	Example
		2,200 - 600 = 1,600
2,238 − 599		*two thousand, two hundred thirty-eight* minus *five hundred ninety-nine*
		1,639
		2,200 − 600 = 1,600

Addition Problem	Vocabulary Word	Example
		5,900 + 700 = 6,600
5,877 + 673		*five thousand, eight hundred seventy-seven* plus *six hundred seventy-three*
		6,550
		5,900 + 700 = 6,600

MY Vocabulary Cards

Lesson 2–1

Associative Property of Addition

$$(13 + 10) + 4 = 13 + (10 + 4)$$

Lesson 2–1

Commutative Property of Addition

$$12 + 15 = 15 + 12$$

Lesson 2–9

equation

$$a + 2 = 5; \ 7 - b = 4$$

Lesson 2–1

Identity Property of Addition

$$0 + 18 = 18 \quad | \quad 18 + 0 = 18$$

Lesson 2–6

minuend

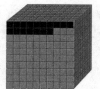

$$1,000 - 17 = 983$$

Lesson 2–6

subtrahend

$$1,000 - 17 = 983$$

Lesson 2–1

unknown

$$150 + 300 - 200 = \ ?$$

Lesson 2–9

variable

$$150 + 300 - k = 250$$

Ideas for Use

- Draw examples for each card. Make drawings that are different from what is shown on each card.

- Practice your penmanship! Write each word in cursive.

The property which states that the order in which two numbers are *added* does not change the *sum*.

Imagine you need to shop for many items. Describe how this property could be used when estimating the sum.

The property which states that the grouping of the *addends* does not change the *sum*.

Associate can mean "to combine or come together." Explain how this helps you understand this property.

For any number, zero plus that number is the number.

Look in the dictionary for a meaning of *identity*. Write a sentence using that meaning.

A sentence that contains an equals sign (=), showing that two expressions are equal.

Explain how an equation is different than an expression.

A number subtracted from another number.

Write a tip to help you remember which number is the minuend and which is the subtrahend.

The number from which another number is subtracted.

What word from this chapter is also a part of a subtraction equation, other than minuend?

A letter or symbol used to represent an unknown quantity.

Variable can mean "changeable." Explain how this relates to the math meaning of *variable*.

An amount that has not been identified.

The prefix *un-* means "not." Name two other words with this prefix.

FOLDABLES® Follow the steps on the back to make your Foldable.

Subtraction Equation	Check with Addition
2,161 − 125 = 2,036	2,036 + 125 = 2,161
7,013 − 1,692 = 5,321	5,321 + 1,692 = _____
____ − ____ = ____	____ + ____ = ____

Addition Equation

Check with Subtraction

$2{,}036 + 125 = 2{,}161$

$2{,}161 - 125 = 2{,}036$

$5{,}321 + 1{,}692 = 7{,}013$

$7{,}013 - 1{,}692 = \underline{\hspace{1cm}}$

$\underline{\hspace{1cm}} + \underline{\hspace{1cm}} = \underline{\hspace{1cm}}$

$\underline{\hspace{1cm}} - \underline{\hspace{1cm}} = \underline{\hspace{1cm}}$

Addition Properties and Subtraction Rules

Lesson 1

ESSENTIAL QUESTION
What strategies can I use to add or subtract?

Addition properties can be used to help solve addition problems.

Math in My World

Watch Tutor

Example 1

Carlos is buying the items shown. Does the order in which the musical instruments are scanned change the total cost?

$10 + $20 = $20 + $10

$ [ ] = $ [ ]

The order in which the instruments are scanned does not change the total cost. This is the Commutative Property of Addition.

Key Concept Addition Properties

Words	**Commutative Property of Addition** The order in which numbers are added does not change the sum.
Examples	4 + 1 = 5 1 + 4 = 5
Words	**Associative Property of Addition** The way in which numbers are grouped when added does not change the sum.
Examples	(5 + 2) + 3 5 + (2 + 3) 7 + 3 5 + 5 10 10
Words	**Identity Property of Addition** The sum of any number and 0 is the number.
Examples	8 + 0 = 8 0 + 8 = 8

Parentheses ( ) show which numbers are added first.

Example 2

There were 16 people at the pool on Saturday. There were no people at the pool on Sunday. How many people were there on Saturday and Sunday?

_____ + _____ = _____ This is the _____ Property of Addition.

So, there were _____ people at the pool on Saturday and Sunday.

You can use properties and rules to find the **unknown**, or missing number, in a number sentence.

Example 3

Find the unknown in 10 − ■ = 10.

When you subtract 0 from any number, the result is the number.

So, the unknown is _____.

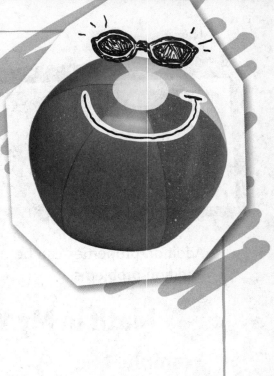

Key Concept   Subtraction Rules

Words	When you subtract 0 from any number, the result is the number.
Examples	22 − 0 = 22 14 − 0 = 14
Words	When you subtract any number from itself, the result is 0.
Examples	16 − 16 = 0 20 − 20 = 0

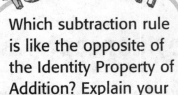

Talk MATH

Which subtraction rule is like the opposite of the Identity Property of Addition? Explain your reasoning.

Guided Practice

Find each unknown. Draw a line to identify the property or rule used.

1. 19 − ■ = 19

   ■ = _____

2. (5 + ■) + 2 = 5 + (9 + 2)

   ■ = _____

3. 74 + 68 = ■ + 74

   ■ = _____

- Commutative Property of Addition

- Associative Property of Addition

- When you subtract 0 from any number, the result is the number.

Independent Practice

Algebra Find each unknown. Write each property or rule that is used.

4. $(\blacksquare + 8) + 7 = 9 + (8 + 7)$

$\blacksquare = $ _____

_____

5. $14 + 13 = 13 + \blacksquare$

$\blacksquare = $ _____

_____

6. $\blacksquare + 0 = 19$

$\blacksquare = $ _____

_____

7. $25 - \blacksquare = 0$

$\blacksquare = $ _____

_____

8. $17 + (11 + 18) = (17 + \blacksquare) + 18$

$\blacksquare = $ _____

_____

9. $37 - \blacksquare = 37$

$\blacksquare = $ _____

_____

Use the properties of addition to add.

10. $17 + 0 = $ _____

11. $(22 + 35) + 15 = $ _____

12. $16 + 22 = $ _____

13. $0 + 47 = $ _____

14. $19 + (61 + 15) = $ _____

15. $27 + (43 + 16) = $ _____

16. $23 + 74 = $ _____

17. $(24 + 24) + 16 = $ _____

18. $0 + 83 = $ _____

19. $25 + (35 + 19) = $ _____

Problem Solving

20. Paco has 75 minutes before he needs to get ready for baseball practice. He cleans his room for 40 minutes and reads for 35 minutes. How much time will he have left before his baseball practice? Explain.

_____

_____

My Work!

21. Mathematical **PRACTICE 7** **Identify Structure** Chloe ate 10 grapes and 5 crackers. Layla ate 5 grapes and 10 crackers. Who ate more food items? Write a number sentence. Then identify the property or rule used.

_____

_____

_____

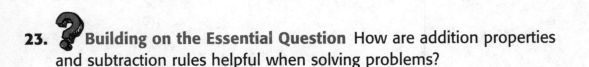

HOT Problems

22. Mathematical **PRACTICE 2** **Use Number Sense** $(23 + \blacksquare) + 19 = 23 + (\blacksquare + 19)$

Can any number complete the number sentence? Explain.

_____

_____

_____

23. ❓ **Building on the Essential Question** How are addition properties and subtraction rules helpful when solving problems?

_____

_____

_____

MY Homework

Homework Helper eHelp

Need help? connectED.mcgraw-hill.com

Add (44 + 18) + 22 mentally.

Use the Associative Property of Addition to make these numbers easier to add. The way in which numbers are grouped when added does not affect the sum.

$(44 + 18) + 22 = 44 + (18 + 22)$ ◄─ Find 18 + 22 first.

$\qquad\qquad\qquad = 44 + 40$

$\qquad\qquad\qquad = 84$

So, $44 + 18 + 22 = 84$.

Practice

Complete each number sentence. Identify the property or rule used.

1. $85 + 0 =$ _____

_____

2. $96 + 13 = 13 +$ _____

_____

3. _____ $- 0 = 37$

_____

_____

4. $(15 + 23) + 7 = 15 + ($ _____ $+ 7)$

_____

Problem Solving

5. While bird watching, Gabrielle saw 6 robins and 3 blue jays. Chase saw 3 robins and 6 blue jays. Who saw more birds? Tell which property you used.

_____

_____

6. Mathematical 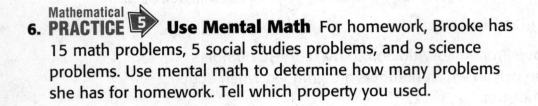 **PRACTICE** **Use Mental Math** For homework, Brooke has 15 math problems, 5 social studies problems, and 9 science problems. Use mental math to determine how many problems she has for homework. Tell which property you used.

_____

7. A soccer team scored 2 goals in the first half. If they won the game by a score of 2 goals to 1 goal, how many goals did they score in the second half? Tell which property you used.

_____

Vocabulary Check

Write a number sentence that demonstrates each property.

8. Commutative Property of Addition _____

9. Associative Property of Addition _____

10. Identity Property of Addition _____

Test Practice

11. Which number sentence represents the Commutative Property of Addition?

  Ⓐ $357 + 0 = 357$ Ⓒ $36 + 14 = 14 + 36$

  Ⓑ $(7 + 19) + 3 = 7 + (19 + 3)$ Ⓓ $79 - 79 = 0$

Number and Operations in Base Ten
4.NBT.4, 4.OA.5

CCSS

Addition and Subtraction Patterns

Lesson 2

ESSENTIAL QUESTION
What strategies can I use to add or subtract?

Math in My World

Tutor

Example 1

On Friday, 1,323 people saw the new movie at a local theater. On Saturday, 1,000 more people saw the new movie. On Sunday, 100 less people saw the movie than on Saturday. How many people saw the movie each day?

1 Cut 4 strips of paper. Place each strip of paper on each of the columns shown. Cover the numbers.

1,000	100	10	1
1,000	100	10	1
1,000	100	10	1

2 Slide the strips of paper up to show 1 thousand, 3 hundreds, 2 tens, and 3 ones.

3 Slide the paper up one row in the thousands column to show the number of people who saw the movie on Saturday.

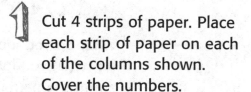

	thousands	hundreds	tens	ones
Friday →	1	3	2	3
Saturday →	2	3	2	3
Sunday →				

4 Slide the paper down one row in the hundreds column to show the number of people who saw the movie on Sunday. Write this number in the place-value chart.

So, _____ people saw the movie on Saturday and _____ saw the movie on Sunday.

Example 2

Miss Starcher drew a puzzle on the whiteboard. The puzzle contains a pattern. Solve her puzzle by filling in the two blank boxes.

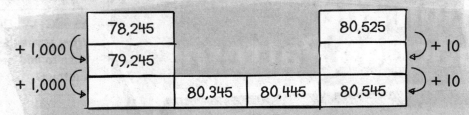

78,245			80,525
79,245			
	80,345	80,445	80,545

+ 1,000
+ 1,000
+ 10
+ 10

1 Each number in the first column

is _____ more than the number

in the row above it.

So, 79,245 + 1,000 = _____ .

2 Each number in the last column is

_____ more than the number

in the row above it.

So, 80,525 + 10 = _____ .

Check

Each number in the last row is _____ more than the number before it.

Since _____ + 100 = 80,345, the answer in the first column

is correct.

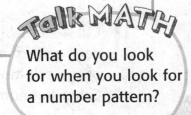

Talk MATH

What do you look for when you look for a number pattern?

Guided Practice

Write each number.

1. 1,000 more than 3,872

2. 10 less than 221

_____

_____

Complete the table.

	Start	End	Change
3.	37,828	38,828	
4.	830,174		100,000 less

Complete each number sentence.

5. 36,525 + _____ = 36,625

6. 98,264 − _____ = 88,264

Independent Practice

Write each number.

7. 100 less than 37,972 _____

8. 10,000 more than 374 _____

9. 10 more than 45,301 _____

10. 1 more than 12,349 _____

11. 10,000 less than 12,846 _____

12. 1,000 more than 91,928 _____

13. 1 less than 37,937 _____

14. 1,000 less than 82,402 _____

Complete the table.

	Start	End	Change
15.	28,192		100 less
16.	8,392	8,402	
17.	521,457	520,457	
18.	51,183		1 more

Complete each number sentence.

19. 45,311 + _____ = 46,311

20. 28,400 − _____ = 28,390

21. 89,420 − _____ = 89,320

22. 84,552 + _____ = 94,552

23. 6,339 + _____ = 6,340

24. 3,014 + _____ = 13,014

Identify and complete each number pattern.

25.

8,901	8,911	8,921			more

26.

	969,987	979,987		999,987	more

27.

56,789		56,589	56,489	56,389	

28.

42,578			42,608	42,618	

Problem Solving

29. Go up the ladder. Write the resulting number on each rung.

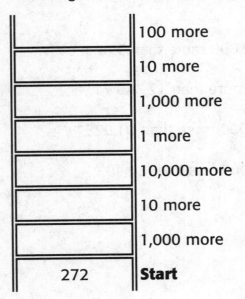

	100 more
	10 more
	1,000 more
	1 more
	10,000 more
	10 more
	1,000 more
272	**Start**

30. Go down the ladder. Write the resulting number on each rung.

12,393	**Start**
	10,000 less
	100 less
	100 less
	1,000 less
	1,000 less
	100 less
	10 less

HOT Problems

31. **Mathematical PRACTICE 3** **Find The Error** Gary completed this number pattern. Find and correct his mistake.

27,389; 26,389; 25,389; 23,389; 24,389

_____

_____

32. **Mathematical PRACTICE 2** **Use Number Sense** Beverages at the Quick Mart increase in price. If this pattern continues, what would be the price of the gallon of milk?

79¢ $1.79 $2.79   [ ]

33. **Building on the Essential Question** Why do we study patterns in mathematics?

_____

_____

Name ....

MY Homework

Homework Helper eHelp

Need help? 🖱 connectED.mcgraw-hill.com

Identify and complete the number pattern.

12,345; 13,345; _____ ; 15,345; 16,345

Look to see how each number is different from the number before it.

Write the first number.   12,345

Write the second number.   13,345

> The place that changed in value is the thousands.

The pattern is 1,000 more or + 1,000.
So, to complete the pattern, add 1,000.

$$\begin{array}{r} 13,345 \\ +\ 1,000 \\ \hline 14,345 \end{array}$$

Check to make sure the pattern continues.

12,345   13,345   14,345   15,345   16,345

+ 1,000   + 1,000   + 1,000   + 1,000

So, the missing number is 14,345.

Practice

Write each number.

1. 100 less than 877

2. 10,000 more than 6,310

3. 10 more than 1,146

4. 1,000 less than 9,052

5. 1,000 more than 37,542

6. 10 less than 2,727

Complete each number sentence.

7. $1,100 + \underline{\hspace{1.5cm}} = 1,200$

8. $40,619 - \underline{\hspace{1.5cm}} = 39,619$

9. $63,088 - \underline{\hspace{1.5cm}} = 53,088$

10. $4,514 + \underline{\hspace{1.5cm}} = 4,524$

Complete each pattern.

11. 7,213; \underline{\hspace{1.5cm}}; 7,413; 7,513

12. 32,877; 42,877; 52,877; \underline{\hspace{1.5cm}}

13. 967, 957, \underline{\hspace{1.5cm}}, 937

14. 3,222; \underline{\hspace{1.5cm}}; 3,220; 3,219

Problem Solving

15. Cady had 435 marbles. One week she used her allowance to buy more, and she had 445 marbles. The next week, she bought more and had 455 marbles. Circle the correct pattern.

   100 more   10 more   10 less   100 less

16. **Mathematical PRACTICE** **8** **Look for a Pattern** Lito keeps track of books in a warehouse. Every month he updates a chart that shows the number of books in stock. For one title, his chart shows 54,350; 44,350; 34,350. If the pattern continues, what number will Lito's chart show next month?

_____

17. Angela uses a large pitcher to water her plants. When the pitcher is full, it holds 384 ounces. Angela gives each plant 100 ounces of water. How much water is left in the pitcher after Angela waters two plants?

_____

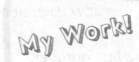

Test Practice

18. Identify the number pattern. 21,344; 20,344; 19,344

   Ⓐ 10 less   Ⓒ 1,000 less

   Ⓑ 10,000 less   Ⓓ 100 less

Add and Subtract Mentally

Lesson 3

ESSENTIAL QUESTION
What strategies can I use to add or subtract?

To mentally add or subtract larger numbers, you can take and give to make one number end in a ten, hundred, or thousand.

 Math in My World Tutor

Example 1

The table shows the number of instruments sold at an instrument store. What is the total number of guitars and trumpets that was sold?

Instruments Sold	
Instrument	**Number Sold**
guitar	223
trumpet	67

Find 223 + _____.

Make a ten.

Take from one addend. → $223 + 67$ ← Give to the other addend.

$\quad\quad -3 \quad\ +3$

$\quad\quad 220 \quad 70$

$220 + 70 =$ _____ ← Write a number sentence.

So, there were _____ guitars and trumpets sold.

Example 2

Find 184 − 59.

Make a ten. 59 is close to 60.

Add 1 to 59 to make 60.
↓
$184 − 60 = 124$

Since you subtracted 1 too many, add it back in. $124 + 1 = 125$

So, 184 − 59 = _____.

Example 3

Attendance at a concert was 82,000 people. The following week, there were 76,000 people. How many more people attended the concert the first week?

Find 82,000 − 76,000.

Both numbers have the same greatest place-value position.

The greatest place-value position is ten thousands. →

TTH	TH	H	T	O
8	2	0	0	0
7	6	0	0	0

 First subtract the ten thousands and the thousands.

82 − 76 = _____

 The difference, 6, is in the thousands place.

So, 82,000 − 76,000 = _____.

So, _____ thousand more people attended the first concert.

Guided Practice ✓Check

Make a ten, hundred, or thousand to mentally add.

1. 57 + 58

−☐ + 2
_____
☐ + 60 = _____

So, 57 + 58 = _____.

2. 499 + 77

+ 1 − ☐
_____
500 + ☐ = _____

So, 499 + 77 = _____.

Talk MATH

Look at Exercise 3. Explain why you added 4 to the difference of 104 before writing the final answer.

Use mental math to subtract.

3. 184 − 76

Make a ten. 76 + 4 = _____

184 − _____ = _____

104 + 4 = _____

So, 184 − 76 = _____.

Independent Practice

4. 8,825 − 6,397

Make a hundred.

6,397 + 3 = _____

_____ − 6,400 = _____

2,425 + 3 = _____

So, 8,825 − 6,397 = _____.

5. 684 − 169

Make a ten.

169 + 1 = _____

_____ − 170 = _____

514 + 1 = _____

So, 684 − 169 = _____.

Make a ten, hundred, or thousand to mentally add.

6. 738 + 56 = _____

7. 223 + 728 = _____

8. 6,627 + 3,315 = _____

9. 5,478 + 1,312 = _____

Use mental math to subtract.

10. 7,930 − 4,623 = _____

11. 5,547 − 2,539 = _____

12. 8,329 − 7,218 = _____

13. 3,273 − 1,256 = _____

Subtract. Draw a line to the difference in the second column.

14. 15,000 − 8,000

• 103,000

15. 77,000 − 65,000

• 12 thousand

16. 394,000 − 44,000

• 7 thousand

17. 273,000 − 170,000

• 350,000

Problem Solving

Use mental addition or subtraction to solve.

18. Attendance at a concert was 12,769. The following evening, the attendance was 13,789. How many more people attended the concert on the second night?

_____

Mathematical PRACTICE 2 **Reason** Yolanda is finding 23,567 − 12,458. She added 2 to 12,458 before subtracting. Then she added it in again after she subtracted. Is her method correct? Explain.

19.

_____

_____

_____

₹₹₹ HOT Problems

20. **Mathematical PRACTICE 1** **Make Sense of Problems** Mentally add or subtract to get to the finish line.

Start	3,829	4,829		
				6,729
			16,629	
315,629		115,629	15,629	
315,639				
	415,649			715,649 **Finish** 🏆

21. ❓ **Building on the Essential Question** Why is mental addition and subtraction important when you learn more difficult concepts?

_____

_____

MY Homework

Homework Helper eHelp

Need help? connectED.mcgraw-hill.com

Find 237 + 48.

Make a ten to mentally add.   237 + 48

Give to the other addend. $+3$ -3 Take from one addend.

$$240 + 45 = 285$$

So, 237 + 48 = 285.

Find 752 − 23.

752 − 20   ← Change 23 to 20 by subtracting 3.

752 − 20 = 732

732 − 3 = 729   ← Since you subtracted 3 too few, subtract 3 more from the total.

So, 752 − 23 = 729.

Practice

Make a ten, hundred, or thousand to mentally add.

1. 118 + 203 = _____

2. 549 + 24 = _____

3. 1,198 + 46 = _____

4. 745 + 997 = _____

Use mental math to subtract.

5. 982 − 56 = _____

6. 7,499 − 4,100 = _____

Problem Solving

Use mental math to solve.

7. The cafeteria sells 498 cartons of milk and 246 bottles of juice every day. How many total milk and juice containers are sold each day?

_____

8. **Mathematical PRACTICE 5** **Use Math Tools** Terrell had a chain of 56 paper clips. Some of the paper clips fell off. Now there are 38 paper clips in the chain. How many fell off?

_____

9. Li counts 203 stars in the sky. The next night, she counts 178 stars. How many stars in all did she see on the two nights?

_____

10. There were 132 children at the museum on Saturday. On Sunday, there were 61 children. How many children were there on Saturday and Sunday?

_____

Test Practice

11. Mikah's dog has 39 spots. Jonelle's dog has 85 spots. How many more spots does Jonelle's dog have?

   Ⓐ 45 spots

   Ⓑ 46 spots

   Ⓒ 41 spots

   Ⓓ 44 spots

Name _____

Estimate Sums and Differences

Lesson 4

ESSENTIAL QUESTION
What strategies can I use to add or subtract?

When estimating, you can round to any place value.

 Math in My World Watch Tutor

Example 1

The Central School District needs 5,481 forks and 2,326 spoons for a school function. About how many forks and spoons will they need altogether?

Estimate 5,481 + 2,326. Round to the hundreds place.

Round each number to the nearest hundred. Then add.

5,481 ⟶ (rounds to) ⟶ ☐,☐☐☐

+ 2,326 ⟶ (rounds to) ⟶ + ☐,☐☐☐

☐,☐☐☐

So, 5,481 + 2,326 is about _____.

Example 2 Watch Tutor

Estimate $7,542 − $3,225. Round to the hundreds place.

$7,542 ⟶ (rounds to) ⟶ $ ☐,☐☐☐

− $3,225 ⟶ (rounds to) ⟶ − $ ☐,☐☐☐

$ ☐,☐☐☐

So, $7,542 − $3,225 is about _____.

Online Content at ⟋ connectED.mcgraw-hill.com

Example 3
Tutor

The table shows populations for two cities in Kentucky. About how many more people live in Covington than in Ashland?

Kentucky Populations	
City	**Population**
Ashland	21,510
Covington	42,811

Round each population to the nearest thousand. Then, subtract.

42,811 ⟶ (rounds to) ⟶ ☐ ☐ , ☐ ☐ ☐

− 21,510 ⟶ (rounds to) ⟶ − ☐ ☐ , ☐ ☐ ☐

☐ ☐ , ☐ ☐ ☐

So, Covington has about _____ more people.

Guided Practice
Check

Estimate. Round each number to the given place value.

1. 1,454 + 335; hundreds

_____ + _____ = _____

2. 2,871 + 427; hundreds

_____ + _____ = _____

3. $2,746 − $1,529; tens

_____ − _____ = _____

4. 48,344 − 7,263; thousands

_____ − _____ = _____

Talk MATH

Estimate 829 + 1,560 to the nearest hundred and the nearest thousand.

Independent Practice

Estimate. Round each number to the given place value.

5. $5,238 + $3,420; hundreds

6. $4,127 + $2,666; hundreds

7. 5,342 + 298; hundreds

8. 3,182 + 6,618; hundreds

9. 48,205 + 50,214; thousands

10. $25,497 + $54,088; ten thousands

11. $7,172 − $5,103; hundreds

12. 9,185 − 6,239; thousands

13. 2,647 − 256; hundreds

14. 27,629 − 5,364; thousands

15. $27,986 − $4,521; thousands

16. $47,236 − $20,425; thousands

Problem Solving

The table shows the tallest buildings in the world. Round each height to the nearest hundred. Write a number sentence to solve.

Tallest Buildings in the World		
Building	**Location**	**Height (ft)**
Taipei 101	Taiwan	1,669
Petronas Towers	Malaysia	1,482
Willis Tower	United States	1,450
Jin Mao Building	China	1,381
CITIC Plaza	China	1,282
Shun Hing Square	China	1,259
Empire State Building	United States	1,250

17. About how much taller is the Willis Tower than the Jin Mao Building?

_____

_____

Mathematical
18. **PRACTICE 4 Model Math** Estimate the difference between the height of the Taipei 101 building and the Empire State Building.

_____

19. About how much taller is Petronas Towers than the Empire State Building?

_____

My Work!

HOT Problems

Mathematical
20. **PRACTICE 2 Reason** Write two numbers that when rounded to the thousands place have an estimated sum of 10,000.

_____

21. **? Building on the Essential Question** How do you know if an estimate is reasonable? Explain.

_____

_____

_____

Name ....

MY Homework

Homework Helper

Need help? connectED.mcgraw-hill.com

Estimate 468 + 2,319. Round to the nearest hundred.

$$
\begin{array}{r}
468 \\
+ 2{,}319
\end{array}
\quad
\begin{array}{l}
\text{rounds to} \rightarrow \\
\text{rounds to} \rightarrow
\end{array}
\quad
\begin{array}{r}
500 \\
+ 2{,}300 \\
\hline
2{,}800
\end{array}
$$

So, 468 + 2,319 is about 2,800.

Estimate 55,599 − 22,782. Round to the nearest thousand.

$$
\begin{array}{r}
55{,}599 \\
- 22{,}782
\end{array}
\quad
\begin{array}{l}
\text{rounds to} \rightarrow \\
\text{rounds to} \rightarrow
\end{array}
\quad
\begin{array}{r}
56{,}000 \\
- 23{,}000 \\
\hline
33{,}000
\end{array}
$$

So, 51,599 − 22,782 is about 33,000.

Practice

Estimate. Round each number to the nearest hundred.

1. 7,392 — rounds to ▸ ⬚,⬚⬚⬚
    + 4,112 — rounds to ▸ + ⬚,⬚⬚⬚
    ⬚⬚,⬚⬚⬚

2. 8,752 — rounds to ▸ ⬚,⬚⬚⬚
    − 3,269 — rounds to ▸ − ⬚,⬚⬚⬚
    ⬚,⬚⬚⬚

Estimate. Round each number to the nearest thousand.

3. $5,486 + $8,602

4. 95,438 − 62,804

Problem Solving

**Estimate. Round each number
to the nearest hundred.**

5. A total of 2,691 people attended the school
play. A total of 1,521 people attended the band
concert. About how many more people attended
the play than the concert?

_____

**Estimate. Round each number
to the nearest thousand.**

6. The highest point in Texas, Guadalupe Peak, is
8,749 feet high. The highest point in California,
Mount Whitney, is 14,497 feet high. About how
much higher is Mount Whitney than Guadalupe
Peak?

_____

7. **Mathematical**
PRACTICE **2** **Use Number Sense** Maria's
school raised $23,240 in magazine sales and
Cole's school raised $16,502. About how much
more money did Maria's school raise?

_____

Test Practice

8. Which is the correct estimate for 63,621 − 41,589
rounded to the nearest hundred?

   Ⓐ 22,040

   Ⓑ 22,000

   Ⓒ 20,000

   Ⓓ 22,032

Check My Progress

Vocabulary Check

1. Each mouse is using a property of addition. Find each unknown. Draw lines through the maze to help each mouse find the cheese with the property that matches its addition sentence(s).

56 + 13 = _____
13 + 56 = _____

Commutative   Identity   Associative

42 + 38 = _____
38 + 42 = _____

0 + 63 = _____

(62 + 18) + 45 = _____
62 + (18 + 45) = _____

78 + 0 = _____

(24 + 14) + 53 = _____
24 + (14 + 53) = _____

Concept Check

Write each number.

2. 1,000 less than 49,737

3. 10,000 more than 53,502

_____   _____

4. Sarah and her mom are at the mall. They buy a shirt that costs $16, a belt that costs $8, and a dress that costs $22. To find the total cost, Sarah adds $16 and $8, and then adds that sum to $22. Her mom adds $16 to the sum of $8 and $22. What property of addition are they using to find the total cost? What is the total cost?

_____

5. Mr. Cleff wants to buy the following instruments for band class.

About how much money will he need to buy the instruments above?

_____

Test Practice

6. Which number sentence can be used to estimate 3,401 + 8,342?

Ⓐ 3,000 + 8,000 = 11,000

Ⓑ 3,000 + 9,000 = 12,000

Ⓒ 4,000 + 8,000 = 12,000

Ⓓ 4,000 + 9,000 = 13,000

Add Whole Numbers

Lesson 5

ESSENTIAL QUESTION
What strategies can I use to add or subtract?

When you add numbers, it might be necessary to regroup.

 ## Math in My World Watch Tutor

I got it!

Example 1

There were 6,824 baseball tickets sold last week. This week, there were 349 baseball tickets sold. How many tickets were sold altogether?

Find 6,824 + 349.

 Add ones.

4 + 9 = 13

Regroup 13 ones as 1 ten and 3 ones.

 Add tens.

1 + 2 + 4 = 7

 Add hundreds.

8 + 3 = 11

Regroup 11 hundreds as 1 thousand and 1 hundred.

 Add thousands.

1 + 6 = 7

	6 ,	8	2	4
+		3	4	9
	,			

So, _____ tickets were sold altogether.

Check for Reasonableness The estimate is _____ .

Since _____ is close to the estimate, the answer is reasonable.

Online Content at connectED.mcgraw-hill.com

Example 2

Ticket sales for a play are shown in the table. What were the total sales?

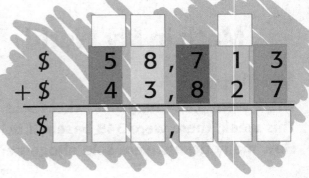

Ticket Sales	
Day	**Amount**
Saturday	$58,713
Sunday	$43,827

Estimate

$58,713	rounds to	$60,000
+ $43,827	rounds to	+ $40,000
		$100,000

 Add ones.

$3 + 7 = 10$

Regroup 10 ones as 1 ten and 0 ones.

 Add tens.

$1 + 1 + 2 = 4$

 Add hundreds.

$7 + 8 = 15$

Regroup 15 hundreds as 1 thousand and 5 hundreds.

 Add thousands.

$1 + 8 + 3 = 12$

Regroup 12 thousands as 1 ten thousand and 2 thousands.

 **Add ten thousands.**

$1 + 5 + 4 = 10$

Place the $ symbol in front of the sum.

So, the total ticket sales were _____.

Talk MATH

Explain why it is important to line up digits in numbers when you add.

Guided Practice ✓Check

Add. Estimate to check your work.

1. $2,961
    + $4,205

2. 29,380
    + 10,225

Independent Practice

Add. Estimate to check your work.

3. 8,346
    + 7,208

4. $23,824
    + $ 7,346

5. 82,828
    + 4,789

6. $37,178
    + $82,370

7. $693,782
    + $ 47,816

8. 743,980
    + 211,315

9. 254,671
    + 381,366

10. $15,789
    + $22,503

11. 56,772
    + 29,428

Add. Use the place-value chart to help set up the problem.

12. 17,599 + 72,682 = _____

Thousands			Ones		
hundreds	tens	ones	hundreds	tens	ones

Problem Solving

13. There are 4,585 students who rode the bus to school today. There were 3,369 students who came to school another way. How many students were there in all at the school?

14. Mathematical  **PRACTICE** **Explain to a Friend** Becky's mom wants to buy a new TV that costs $1,500 and a DVD player that costs $300. She has $2,000. If she buys groceries for $150, will she have enough money for the TV and the DVD player? Explain to a friend.

15. Mr. Russo's class is collecting bottles to recycle. The class collected 1,146 bottles in March and 2,555 bottles in April. How many bottles were collected?

HOT Problems

16. Mathematical **PRACTICE** **Make Sense of Problems** Write two 5-digit addends whose sum would give an estimate of 60,000.

17. **Building on the Essential Question** Explain why an addition problem that has 4-digit addends could have a 5-digit sum.

Name _____

MY Homework

Homework Helper

Need help? connectED.mcgraw-hill.com

Find 32,866 + 7,375.

Estimate 32,866   rounds to   33,000
        + 7,375   rounds to   + 7,000
                                      40,000

1 Add ones.
$6 + 5 = 11$
Regroup 11 ones as 1 ten and 1 one.

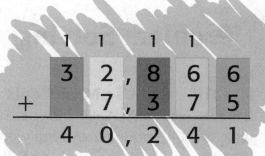

2 Add tens.
$1 + 6 + 7 = 14$
Regroup 14 tens as 1 hundred and 4 tens.

3 Add hundreds.
$1 + 8 + 3 = 12$
Regroup 12 hundreds as 1 thousand and 2 hundreds.

4 Add thousands.
$1 + 2 + 7 = 10$
Regroup 10 thousands as 1 ten thousand and 0 thousands.

5 Add ten thousands.
$1 + 3 = 4$

So, 32,866 + 7,375 = 40,241.

40,241 is close to the estimate of 40,000. The answer is reasonable.

Practice

Add. Estimate to check your work.

1.   5,239
   + 2,794

2. $4,189
   + $5,432

3.   169,748
   + 355,470

4.   452,903
   + 318,766

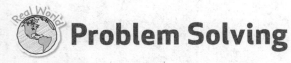 **Problem Solving**

5. **Mathematical PRACTICE** 5 **Use Math Tools** A zoo has two elephants, Sally and Joe. Sally weighs 7,645 pounds, and Joe weighs 12,479 pounds. How much do Sally and Joe weigh in all?

_____

My Work!

6. At a library, 1,324 children's books and 1,510 fiction books were checked out. How many books were checked out at the library?

_____

Test Practice

7. Find the unknown in $45,209 + $31,854 = ■.

Ⓐ $76,063   Ⓒ $77,053

Ⓑ $77,163   Ⓓ $77,063

Subtract Whole Numbers

Lesson 6

ESSENTIAL QUESTION
What strategies can I use to add or subtract?

Subtraction of whole numbers is similar to addition of whole numbers because you might need to regroup.

 ## Math in My World Watch Tutor

Example 1

The Trevino family is moving to a new city. They have driven 957 miles out of the 3,214 miles that they need to drive. How many more miles do they need to drive?

Find 3,214 − 957.

 Subtract ones.
Regroup a ten as 10 ones.
10 ones + 4 ones = 14 ones
14 ones − 7 ones = _____ ones

 Subtract tens.
Regroup a hundred as _____ tens.
10 tens + 0 tens = 10 tens
10 tens − 5 tens = _____ tens

 Subtract hundreds.
Regroup a thousand as _____ hundreds.
10 hundreds + 1 hundred = 11 hundreds
11 hundreds − 9 hundreds = _____ hundreds

 Subtract thousands.
2 thousands − 0 thousands = _____ thousands

So, 3,214 − 957 = _____. The Trevino family

needs to drive _____ more miles.

$$\begin{array}{r} 3,2\ 1\ 4 \\ -\ \ \ 9\ 5\ 7 \\ \hline , \end{array}$$

Online Content at connectED.mcgraw-hill.com

The **minuend** is the first number in a subtraction sentence from which a second number is to be subtracted. The **subtrahend** is the number that is subtracted.

Example 2

The band has raised **$1,345** toward new equipment. If the goal is to raise **$4,275**, how much money must still be raised?

Estimate $4,275   (rounds to) → $4,300

          − $1,345   (rounds to) → $1,300

                            $3,000

 Subtract ones.
$5 - 5 = 0$

 Subtract tens.
$7 - 4 = 3$

 Subtract hundreds.
Regroup a thousand as 10 hundreds.
$12 - 3 = 9$

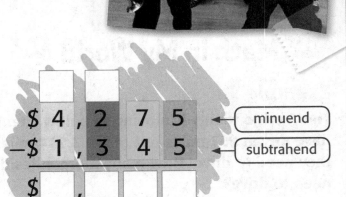

| minuend → | $4,275 |
| subtrahend → | −$1,345 |

 Subtract thousands.
$3 - 1 = 2$

So, the band still needs to raise _____ .

Check for Reasonableness You can use addition to check your subtraction.

   4,275 ⟶ 2,930
− 1,345 ⟶ + 1,345
   2,930 ⟶ 4,275

The answer is correct and close to the estimate.

Talk MATH

Explain how to check the answer to a subtraction problem by using addition.

Guided Practice

Subtract. Use addition or estimation to check.

1.   2,962
     −   845

2. $4,785
     − $2,293

Independent Practice

Subtract. Use addition or estimation to check.

3. 8,845
     − 627

4. $5,751
     − $4,824

5. $8,327
     − $5,709

6. 39,536
     − 18,698

7. 847,311
     − 562,530

8. 93,458
     − 21,649

9. 78,215
     − 56,827

10. $18,345
     − $14,400

11. 629,843
     − 216,954

Subtract. Use addition or estimation to check. Use the place-value chart to set up the problem.

12. 961,344 − 345,822 = _____

Thousands			Ones		
hundreds	tens	ones	hundreds	tens	ones

13. Do you prefer to use addition or estimation to check? Explain.

_____

_____

Problem Solving

14. **Mathematical PRACTICE 5** **Use Math Tools** There are a total of 1,569 tickets for a concert. On the first day of sales, 875 tickets were sold. The following day an additional 213 tickets were sold. How many tickets are still available?

_____

15. A mountain is 29,135 feet tall. From base camp at 17,600 feet, a climber hiked 2,300 feet. How much farther does the climber have before reaching the top of the mountain?

_____

16. John Adams was born in 1732 and became President in 1797. Harry S. Truman was born in 1884 and became President in 1945. Who was older when he became President?

_____

HOT Problems

17. **Mathematical PRACTICE 2** **Use Number Sense** Circle the subtraction problem that does not require regrouping. Explain.

67,457	71,639	89,584	95,947
− 40,724	− 39,607	− 57,372	− 26,377

_____

_____

18. **Building on the Essential Question** Why is it important to line up the digits in each place-value position when subtracting?

_____

_____

_____

Name ....

MY Homework

Homework Helper Need help? connectED.mcgraw-hill.com

Find 6,325 − 2,841.

Estimate Round to the nearest thousand. 6,000 − 3,000 = 3,000

 Subtract ones.

 Subtract tens.
Regroup a hundred as 10 tens.

 Subtract hundreds.
Regroup a thousand as ten hundreds.

 Subtract thousands.

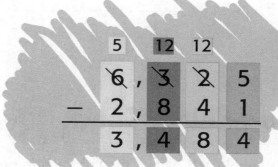

So, 6,325 − 2,841 = 3,484.

Check

Use addition to check.

```
  6,325          3,484
− 2,841        + 2,841
  3,484          6,325
```

So, the answer is reasonable.

Practice

Subtract. Use addition or estimation to check.

1. $6,148
   − $1,575

2. 9,516
   − 7,228

3. 6,637
   − 2,846

4. 33,539
   − 31,649

Problem Solving

5. A minor league baseball team gave out 1,250 free hats. If 2,359 people attended the game, how many did not get a hat?

_____

6. **Mathematical PRACTICE 5** **Use Math Tools** There were 3,515 shirts at the stadium store before the game. After the game, there were 1,396 shirts left. How many were sold during the game?

_____

Vocabulary Check

7. Label each part of the subtraction problem with the correct vocabulary term.

| difference | minuend | subtrahend |

$$
\begin{array}{r}
4{,}178 \\
-\ 535 \\
\hline
3{,}643
\end{array}
$$

Test Practice

8. Find the unknown in $1{,}515 - 1{,}370 = \blacksquare$.

Ⓐ 165

Ⓑ 145

Ⓒ 135

Ⓓ 235

Number and Operations in Base Ten
4.NBT.4

CCSS

Subtract Across Zeros

Lesson 7

ESSENTIAL QUESTION
What strategies can I use to add or subtract?

Sometimes subtraction involves minuends that have zeros.

 Math in My World

Watch ▶ Tutor 💬

Example 1

Each fourth grade class has a goal to collect 5,100 pennies for charity. How many more pennies do the fourth graders in Mr. Blake's class need to reach their goal?

Find 5,100 − 3,520.

Class	Pennies
Mrs. Clark	4,523
Mr. Blake	3,520
Ms. Simms	1,987
Mrs. Stone	2,569

 Subtract ones.
0 ones − 0 ones = 0 ones

 Subtract tens.
Regroup 1 hundred as 10 tens.
10 tens − 2 tens = 8 tens

 Subtract hundreds.
Regroup one thousand as 10 hundreds.
10 hundreds − 5 hundreds = 5 hundreds

 Subtract thousands.
4 thousands − 3 thousands = 1 thousand

$$\begin{array}{r} 5,1\ 0\ 0 \\ -3,5\ 2\ 0 \\ \hline , \end{array}$$

So, Mr. Blake's class needs _____ more pennies.

Check ☐,☐☐☐ The answer is correct.

$$\begin{array}{r} +3,5\ 2\ 0 \\ \hline 5,1\ 0\ 0 \end{array}$$

Online Content at 🖱 **connectED.mcgraw-hill.com**

Example 2 Tutor

There were 30,090 fans at the stadium on Saturday. The next Saturday, there were 22,977 fans. How many more fans were at the stadium on the first Saturday than the second Saturday?

Find 30,090 − 22,977.

 Subtract ones.

Regroup 1 ten as 10 ones.

10 ones − 7 ones = _____ ones

 Subtract tens.

8 tens − 7 tens = _____ ten

 Subtract hundreds.

Regroup 1 ten thousand as 10 thousands.
Regroup 1 thousand as 10 hundreds.

10 hundreds − 9 hundreds = _____ hundred

 Subtract thousands.

9 thousands − 2 thousands = _____ thousands

 Subtract ten thousands.

2 ten thousands − 2 ten thousands = _____ ten thousands

So, 30,090 − 22,977 = _____.

There were _____ more fans than the first Saturday.

```
        □
   □    □ □
  3  0 , 0  9  0
 -2  2 , 9  7  7
   □ , □  □
```

Talk MATH

Explain how to subtract 42,956 from 55,000.

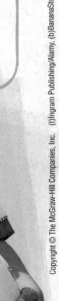

Guided Practice Check

Subtract. Use addition or estimation to check.

1.
```
   2, 0  0  3
 - 1, 1  5  4
   □  □  □
```

2.
```
   $8, 0  0  0
 - $3, 5  0  2
  $□ , □  □  □
```

Name _____

Independent Practice

Subtract. Use addition or estimation to check.

3. 2,040
− 946

4. 7,008
− 2,055

5. 12,050
− 3,162

6. 10,400
− 5,445

7. 46,801
− 5,823

8. 60,032
− 21,833

9. $52,006
− $13,055

10. 600,000
− 28,005

11. 508,200
− 136,118

Subtract. Use addition or estimation to check. Use the place-value chart to set up the problem.

12. 900,000 − 31,650 = _____

Thousands			Ones		
hundreds	tens	ones	hundreds	tens	ones

Problem Solving

For Exercises 13 and 14, use the table which shows the distance between New York City and five other cities around the world.

City	Miles
Jakarta, Indonesia	10,053
London, England	3,471
Mexico City, Mexico	2,086
Munich, Germany	4,042
Paris, France	3,635

13. How many more miles is it to travel to Jakarta than to London?

_____

14. How many more miles is it to travel to Munich than to Paris?

_____

15. **Mathematical PRACTICE 5** **Use Math Tools** Trent earned 4,005 points in a video game. His brother earned 2,375 points in the same game. How many more points did Trent earn than his brother?

_____

HOT Problems

16. **Mathematical PRACTICE 1** **Plan Your Solution** Identify a number that results in a 4-digit number when 156,350 is subtracted from it.

_____

17. **Building on the Essential Question** How does understanding place value help you to subtract across zeros?

_____

_____

_____

Name ....

MY Homework

Homework Helper

Need help? connectED.mcgraw-hill.com

Find 10,200 − 4,795.

 Subtract ones.
Regroup a hundred as 10 tens.
Regroup a ten as 10 ones.

 Subtract tens.

Subtract hundreds.
Regroup a ten thousand as 10 thousands.
Regroup a thousand as 10 hundreds.

Subtract thousands.

Subtract ten thousands.
0 ten thousands − 0 ten thousands = 0 ten thousands.

So, 10,200 − 4,795 is 5,405.

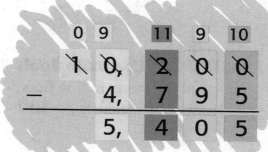

Practice

Subtract. Use addition or estimation to check.

1. 4,000
    − 1,731

2. 3,300
    − 1,892

3. 8,000
    − 6,313

4. $14,000
    − $10,892

5. If 700 tickets were sold for a concert and only 587 people attended, how many people bought a ticket but did not attend?

_____

6. The Amazon River, in South America, is 4,000 miles long. The Snake River, in the northwestern United States, is 1,038 miles long. How much longer is the Amazon River than the Snake River?

_____

7. Mathematical
PRACTICE 5 **Use Math Tools** There are 6,000 products at the store. In one hour, 425 products are sold. How many products are left?

_____

8. A field of corn has 2,000 insects. Only 497 are eating the corn. How many insects are not eating the corn?

_____

Test Practice

9. Logan has a gift card for $200. He spends $45 on Monday and $61 on Tuesday. How much money is left on his gift card?

Ⓐ $94

Ⓑ $106

Ⓒ $104

Ⓓ $139

My Work!

Check My Progress

Vocabulary Check

1. Help each butterfly find its flower by drawing lines to match each vocabulary word with its definition.

minuend

subtrahend

difference

The first number in a subtraction sentence from which a second number is to be subtracted.

The answer to a subtraction problem.

The number that is subtracted in a subtraction problem.

Concept Check

Add. Estimate to check your work.

2. $3,618
    + $2,956

3. 36,847
    + 14,268

4. 529,318
    + 231,937

Subtract. Use addition or estimation to check.

5. 5,428
    − 725

6. $90,000
    − $24,074

7. 836,422
    − 145,742

Problem Solving

8. Jabar traveled 3,052 miles last year. His brother traveled 5,294 miles. How many miles did they travel in all?

_____

9. There were 15,292 concert tickets sold last year. There were 26,935 concert tickets sold this year. How many more tickets were sold this year than last year?

_____

10. A store wants to earn $100,000 this year. So far, they have earned $82,052. How much more money will they have to earn to meet their goal?

_____

11. A book has 31,225 words. A shorter book has 24,893 words. How many more words does the longer book have than the shorter book?

_____

Test Practice

12. Find the unknown.

$$24{,}378 + 12{,}489 = p$$

Ⓐ $p = 36{,}867$

Ⓑ $p = 36{,}757$

Ⓒ $p = 12{,}111$

Ⓓ $p = 11{,}889$

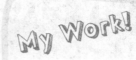

My Work!

Problem-Solving Investigation

STRATEGY: Draw a Diagram

Lesson 8

ESSENTIAL QUESTION
What strategies can
I use to add or subtract?

Learn the Strategy

Keith's summer camp is going to build some tree houses. They will need $2,492 for tools, and $12,607 for wood. How much money do they need to build the tree houses?

1 Understand

What facts do I know?

Tools cost $_____ . Wood costs $_____ .

What do I need to find?
Find how much money is needed to build the tree houses.

2 Plan

I can draw a bar diagram and add to find the sum.

3 Solve

The diagram shows each part that is needed. Add to find the total.

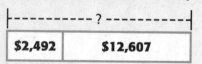

?	
$2,492	$12,607

$$
\begin{array}{r}
\$2,492 \\
+\ \$12,607 \\
\hline
\$\ \square\square,\square\square\square
\end{array}
$$

So, a total of _____ is needed to build the tree houses.

4 Check

Does your answer make sense? Explain.

_____

Practice the Strategy

East School District has 52,672 students. West School District has 34,089 students. How many more students are there in East School District than in West School District?

1 Understand

What facts do I know?

_____

_____

_____

What do I need to find?

_____

_____

_____

2 Plan

_____

_____

3 Solve

4 Check

Does your answer make sense? Explain.

_____

Name _____

Apply the Strategy

Solve each problem by drawing a diagram.

1. **Mathematical PRACTICE** 4 **Model Math** Twickenham Stadium, in England, can seat 82,000 people. If there are 49,837 people seated in the stadium, how many more people can be seated in the stadium?

2. A bakery uses ten cups of butter and ten eggs for a recipe. There are 16,280 Calories in ten cups of butter. Ten eggs have 1,170 Calories. How many more Calories are there in ten cups of butter than in ten eggs?

3. Write and solve a real-world problem that has a sum of 11,982.

_____

_____

_____

_____

Review the Strategies

Use any strategy to solve each problem.
- Use the four-step plan.
- Draw a diagram.

4. Miss Bintel wants to buy a car that costs $35,500. There is a sale on Sunday. If she buys the car on Sunday, she will save $2,499. How much will the car cost on Sunday?

The car will cost _____ on Sunday.

5. Mathematical **PRACTICE** 5 **Use Math Tools** Rick drove 12,363 miles in his new car the first year he owned it. He drove 15,394 miles the second year. How many miles did Rick drive these two years?

_____

My Work!

6. Mrs. Walker has 2,005 recipes to organize. She has organized 962 of them. How many more recipes does she need to organize?

_____

7. A moose weighs 1,820 pounds. A camel weighs 1,521 pounds. How much more does a moose weigh than a camel?

_____

8. Mathematical **PRACTICE** 3 **Find the Error** Macy wants to find the sum of 61,043 and 23,948. Her answer is 37,095. Find and correct her mistake.

_____

_____

MY Homework

Homework Helper eHelp

Need help? connectED.mcgraw-hill.com

On the first day of an audition, 2,731 singers participated. On the second day of an audition, 4,327 singers participated. How many singers participated in all? Use a bar diagram to solve.

1 Understand

What facts do I know?

There were 2,731 singers on the first day.

There were 4,327 singers on the second day.

What do I need to find?

Find how many singers there were in all.

2 Plan

I can draw a bar diagram and add to find the sum.

3 Solve

The diagram shows each part that is needed. Add to find the total.

?	
2,731	**4,327**

$$\begin{array}{r} 2{,}731 \\ +\ 4{,}327 \\ \hline 7{,}058 \end{array}$$

So, a total of 7,058 singers participated.

4 Check

Does your answer make sense? Explain.

2,731 rounds to 3,000. 4,327 rounds to 4,000. 3,000 + 4,000 = 7,000.
7,000 is close to the actual sum of 7,058. So, my answer makes sense.

![Real World] **Problem Solving**

Solve each problem by drawing a diagram.

1. Joseph has 3,124 pieces of paper in his classroom. Emily has 5,229 pieces of paper in her classroom. How many pieces of paper do they have in both classrooms?

_____

2. Brayden sold 2,306 tickets for a school raffle. Connor sold 1,523 tickets for a school raffle. How many more tickets did Brayden sell than Connor?

_____

3. On Saturday, 5,395 people visited a museum. On Sunday, 3,118 people visited a museum. How many people visited altogether on Saturday and Sunday?

_____

Solve Multi-Step Word Problems

Lesson 9

ESSENTIAL QUESTION
What strategies can I use to add or subtract?

You can write an **equation** to help organize and solve multi-step problems. An equation is a sentence that contains an equals sign (=), showing that both sides of the equals sign are equal.

 Math in My World Tutor

Example 1

The music club had $390 in their account. At the concert, they earned $472. Afterwards, they had to pay $75 to rent the stage and $102 for the rental equipment. How much is in their account now?

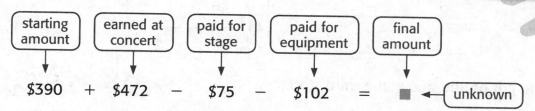

Write an equation.

starting amount	earned at concert	paid for stage	paid for equipment	final amount

$$\$390 \;+\; \$472 \;-\; \$75 \;-\; \$102 \;=\; \blacksquare \;\longleftarrow \text{unknown}$$

When more than one step is involved, add and subtract in order from left to right.

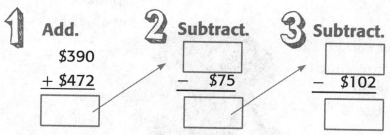

1 Add.

$$\begin{array}{r} \$390 \\ +\ \$472 \\ \hline \end{array}$$

2 Subtract.

$$\begin{array}{r} \\ -\ \$75 \\ \hline \end{array}$$

3 Subtract.

$$\begin{array}{r} \\ -\ \$102 \\ \hline \end{array}$$

So, the account now has _____.

Check An estimate is $400 + $500 − $100 − $100, or _____.

This is close to the actual amount, which is $ _____.
So, the answer is reasonable.

A **variable** is a symbol, often a letter, that is used to represent an unknown, or a number that is not known.

Example 2

Twenty people got on the bus at the first stop. At the second stop, 14 people got off the bus, and 5 people got on the bus. At the third stop, 2 people got off the bus, and some people got on the bus. Then there were 24 people on the bus. How many people got on the bus at the third stop?

Write an equation.
The letter *b* can be used as a variable to represent the unknown.

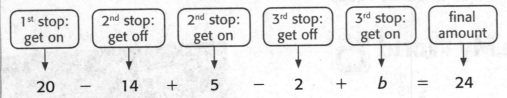

1st stop: get on	2nd stop: get off	2nd stop: get on	3rd stop: get off	3rd stop: get on	final amount

$$20 - 14 + 5 - 2 + b = 24$$

Find the unknown.

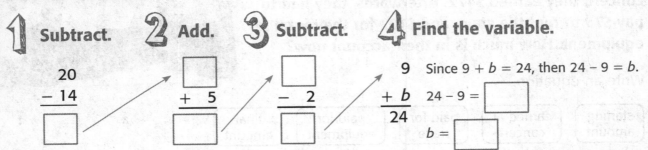

1 Subtract.

20
− 14
⬚

2 Add.

⬚
+ 5
⬚

3 Subtract.

⬚
− 2
⬚

4 Find the variable.

9
+ *b*
24

Since $9 + b = 24$, then $24 - 9 = b$.

$24 - 9 = $ ⬚

$b = $ ⬚

So, ⬚ people got on the bus at the third stop.

Guided Practice

1. **Algebra** Savannah had $15. She earned $20. Then, she bought a gift for $8. How much money does she have left? Write an equation to solve the problem. Use a variable for the unknown.

_____

_____

Talk MATH

Can you use any letter of the alphabet for a variable? Explain.

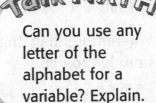

Independent Practice

Algebra **Write an equation to solve each problem.
Use a variable for the unknown.**

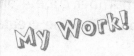

2. Bailey had 75 beads. She used 20 of them on a necklace and 12 of them on a bracelet. Then she bought 25 more beads. How many beads does Bailey have now?

_____

3. Alex had $30. He spent $13 on a game and $5 on a poster. Then, he earned $8 doing chores for a week. How much money does Alex have now?

_____

4. Hunter had 16 jars of paint. He used 2 of them on a painting. He bought 8 more jars. Then, he used some of the jars to make another painting. Now, Hunter has 15 jars of paint. How many jars did he use for the second painting?

_____

5. A restaurant served food to a large party. The manager is listing the total costs, shown below.

Item	Price ($)
chicken	452
pasta	388
salad	150
side dishes	s

The total cost is $1,317. How much did the side dishes cost?

_____

Problem Solving

Use a number cube to complete each number puzzle.

6. Roll a number cube 4 times. Write one number in each box. Find the greatest value of the variable.

$\boxed{} + \boxed{} - \boxed{} + \boxed{} = b$

$b = \underline{\hspace{2cm}}$

7. Roll a number cube 6 times. Write one number in each box. Find the greatest value of the variable.

$\boxed{} + \boxed{} - \boxed{} + \boxed{} + \boxed{} - \boxed{} = y$

$y = \underline{\hspace{2cm}}$

HOT Problems

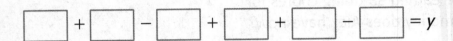

8. **Mathematical PRACTICE 1** **Make Sense of Problems** Victoria had some money in her wallet. She went to the mall and spent $8 on a stuffed animal, $7 on lunch, and $13 on a gift for her mom. Then her brother gave her $10. She bought a book for $15. Now she has $12. How much money did Victoria originally have in her wallet?

9. **Building on the Essential Question** How can I use variables to describe real-world problems? Explain.

MY Homework

Homework Helper eHelp

Need help? connectED.mcgraw-hill.com

The concession stand employees started with $520 in the cash register. They earned $725 at the football game. They had to pay $125 for more popcorn and $65 for more hot chocolate. How much is in the cash register now?

Write an equation.

| starting amount | earned at game | paid for popcorn | paid for hot chocolate | final amount |

$520 + $725 − $125 − $65 = c ← unknown

Add and subtract in order from left to right.

Estimate 520 + 725 − 125 − 65 = c

rounds to

500 + 700 − 100 − 100 = $1,000

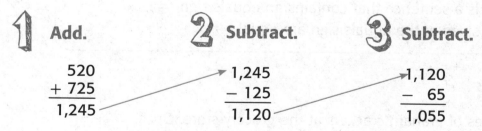

1 Add.

520
+ 725
1,245

2 Subtract.

1,245
− 125
1,120

3 Subtract.

1,120
− 65
1,055

So, the cash register now has $1,055.

Check The estimate is $1,000. This is close to the actual amount, which is $1,055. So, the answer is reasonable.

Problem Solving

Mathematical PRACTICE **2** **Use Algebra** Write an equation to solve each problem. Use a variable for the unknown.

1. Trent's mom gave him $30. He earned another $12 completing chores. Trent spent $15 at the movie theater and $6 on lunch. How much money does Trent have?

2. The cafeteria ordered 400 paper plates. They used 226 at breakfast. They bought 100 more. Then, they used some plates for lunch. Now there are 78 plates. How many plates did they use at lunch?

3. Mia's family has $150 to spend at the beach for the day. It cost them $75 to rent a boat and $35 for lunch. How much money do they have now?

Vocabulary Check

Complete each sentence using the words below.

equation   variable

4. A(n) _____ is a symbol, usually a letter, that is used to represent an unknown, or an amount that has not been identified.

5. A(n) _____ is a sentence that contains an equals sign (=), showing that both sides of the equals sign are equal.

Test Practice

6. There are 367 boxes of graham crackers at the grocery store. On Monday, 126 boxes are sold and on Tuesday, 92 boxes are sold. On Wednesday, 203 more boxes are delivered. How many boxes are there now? Which equation represents this situation?

   (A) $367 + 126 + 92 - 203 = b$ (B) $367 - 126 - 92 + 203 = b$

   (C) $367 + 126 - 92 + 203 = b$ (D) $367 + 126 - 92 + b = 203$

Fluency Practice

Add.

1.   53,035
     + 39,952

2.   94,225
     + 63,236

3.   82,427
     + 37,174

4.   32,472
     + 18,009

5.   72,259
     + 62,905

6.   52,372
     + 17,429

7.   63,141
     + 14,603

8.   20,407
     + 38,692

9.   367,028
     + 52,842

10.   482,952
      + 20,485

11.   137,953
      + 84,037

12.   813,448
      + 92,734

13.   109,374
      + 824,849

14.   372,555
      + 372,555

15.   218,662
      + 741,852

16.   359,751
      + 486,258

17.   118,577
      + 254,009

18.   888,888
      + 102,222

19.   328,805
      + 646,464

20.   335,533
      + 254,009

Name

Subtract.

1.　63,581
−　37,510

2.　72,510
−　62,507

3.　82,404
−　15,840

4.　43,524
−　43,509

5.　42,824
−　29,131

6.　34,108
−　19,888

7.　13,546
−　12,816

8.　45,850
−　29,544

9.　237,482
−　52,851

10.　321,123
−　32,123

11.　137,953
−　84,037

12.　338,200
−　12,658

13.　825,385
−　703,261

14.　651,851
−　215,992

15.　453,166
−　405,556

16.　212,894
−　198,284

17.　489,255
−　281,816

18.　258,914
−　168,876

19.　545,248
−　359,249

20.　605,060
−　488,777

Vocabulary Check

Use the vocabulary words in the word bank to fill in the blanks.

Associative Property of Addition	**Commutative Property of Addition**
equation	**Identity Property of Addition**
minuend	**subtrahend**
unknown	**variable**

1. The _____ states that for any number, zero plus that number is the number.

2. A(n) _____ quantity is an amount whose value needs to be found.

3. The _____ states that the order in which two numbers are added does not change the sum.

4. The first number in a subtraction sentence from which a second number is to be subtracted is the _____.

5. The _____ states that the grouping of the addends does not change the sum.

6. A number that is subtracted from another number is called the

   _____.

7. A(n) _____ is a symbol, usually a letter, that is used to represent an unknown quantity.

8. A(n) _____ is a sentence that contains an equals sign (=), showing that two expressions are equal.

Concept Check ✓

Find each unknown. Write the addition property or subtraction rule that each shows.

9. $35 - \blacksquare = 35$

_____

_____

10. $(16 + 5) + \blacksquare = 16 + (5 + 10)$

_____

_____

11. $83 + 35 = 35 + \blacksquare$

_____

12. $76 + 0 = \blacksquare$

_____

Write each number.

13. 10,000 more than 25,953

_____

14. 1,000 less than 63,035

_____

Make a ten, hundred, or thousand to mentally add.

15. $4,529 + 56 =$ _____

16. $506 + 349 =$ _____

Add. Estimate to check your work.

17.
$$\begin{array}{r} 82,267 \\ +\ 21,037 \\ \hline \end{array}$$

18.
$$\begin{array}{r} 432,901 \\ +\ 177,235 \\ \hline \end{array}$$

19.
$$\begin{array}{r} 206,522 \\ +\ 321,877 \\ \hline \end{array}$$

Subtract. Use addition or estimation to check.

20.
$$\begin{array}{r} \$54,751 \\ -\ \$43,226 \\ \hline \end{array}$$

21.
$$\begin{array}{r} 9,004 \\ -\ 632 \\ \hline \end{array}$$

22.
$$\begin{array}{r} 70,909 \\ -\ 63,485 \\ \hline \end{array}$$

Real World!

Problem Solving

23. Mrs. VanHorn has 1,045 recipes to organize. She has organized 632 of them. How many more recipes does she need to organize?

_____

24. Parker had $32. He earned $10. Then, he bought a video game for $18. How much money does he have left? Write an equation using a variable for the unknown.

_____

_____

25. Abby won 57 points on her first turn in a game. She won 37 more points on her second turn and then lost 19 points on her third turn. She won some more points on her fourth turn. Now Abby has 100 points. How many points did she win on her fourth turn? Write an equation using a variable for the unknown.

_____

_____

Test Practice

26. Rick drove 12,363 miles in his new car the first year he owned it. He drove 15,934 miles the second year. How many miles altogether did Rick drive these two years?

Ⓐ 23,571 miles

Ⓑ 27,297 miles

Ⓒ 28,291 miles

Ⓓ 28,297 miles

Reflect

Use what you learned about addition and subtraction to complete the grapic organizer.

Addition Example

Subtraction Example

ESSENTIAL QUESTION

What strategies can I use to add or subtract?

Vocabulary

Estimate

Reflect on the ESSENTIAL QUESTION Write your answer below.

3 Understand Multiplication and Division

MY World of Fun

Watch a video!

MY Common Core State Standards

 CCSS

Number and Operations in Base Ten

4.NBT.5 Multiply a whole number of up to four digits by a one-digit whole number, and multiply two two-digit numbers, using strategies based on place value and the properties of operations. Illustrate and explain the calculation by using equations, rectangular arrays, and/or area models.

4.NBT.6 Find whole-number quotients and remainders with up to four-digit dividends and one-digit divisors, using strategies based on place value, the properties of operations, and/or the relationship between multiplication and division. Illustrate and explain the calculation by using equations, rectangular arrays, and/or area models.

Operations and Algebraic Thinking *This chapter also addresses these standards:*

4.OA.1 Interpret a multiplication equation as a comparison, e.g., interpret $35 = 5 \times 7$ as a statement that 35 is 5 times as many as 7 and 7 times as many as 5. Represent verbal statements of multiplicative comparisons as multiplication equations.

4.OA.2 Multiply or divide to solve word problems involving multiplicative comparison, e.g., by using drawings and equations with a symbol for the unknown number to represent

the problem, distinguishing multiplicative comparison from additive comparison.

4.OA.4 Find all factor pairs for a whole number in the range of 1–100. Recognize that a whole number is a multiple of each of its factors. Determine whether a given whole number in the range 1–100 is a multiple of a given one-digit number. Determine whether a given whole number in the range 1–100 is prime or composite.

It looks hard, but I think I can get it!

Standards for Mathematical PRACTICE ⬇

1. Make sense of problems and persevere in solving them.
2. Reason abstractly and quantitatively.
3. Construct viable arguments and critique the reasoning of others.
4. Model with mathematics.
5. Use appropriate tools strategically.
6. Attend to precision.
7. Look for and make use of structure.
8. Look for and express regularity in repeated reasoning.

= focused on in this chapter

Name
....

Am I Ready?

 Check ✓ ← Go online to take the Readiness Quiz

Complete each number sentence.

1. $4 + 4 + 4 =$ _____

2. $6 + 6 +$ _____ $+ 6 = 24$

3. $7 + 7 + 7 = 3 \times$ _____

4. $9 + 9 + 9 + 9 =$ _____ $\times 9$

5. Write the multiplication fact modeled by the array to the right.

_____

Circle equal groups of 3 in each array.

6.

★ ★ ★ ★
★ ★ ★ ★
★ ★ ★ ★

7.

((((((
((((((
((((((
((((((

Complete each number pattern.

8. 2, 4, 6, _____, 10, _____, 14

9. 4, 8, 12, _____, 20, 24, _____

10. 5, _____, 15, 20, _____, 30, _____

11. _____, 18, 27, _____, 45, 54, _____

Shade the boxes to show the problems you answered correctly.

How Did I Do? | 1 | 2 | 3 | 4 | 5 | 6 | 7 | 8 | 9 | 10 | 11 |

MY Math Words

Vocab
abc

Review Vocabulary

divide   multiply

Making Connections

Use the review vocabulary to tell which operation(s) to use in each bubble. Solve the word problems.

How many squares are in the array altogether?

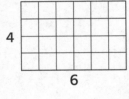

4

6

Five friends arranged 10 total baseball cards into 5 equal groups. How many cards are in each group?

Divide, Multiply,

or Both?

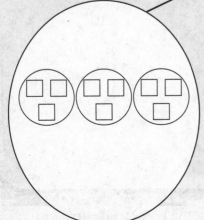

There are 5 baskets. Each basket has 4 pieces of fruit. How many total pieces of fruit are there?

MY Vocabulary Cards

Mathematical
PRACTICE

Copyright © The McGraw-Hill Companies, Inc.

Lesson 3–6

Associative Property of Multiplication

$3 \times (4 \times 6) = (3 \times 4) \times 6$

Lesson 3–5

Commutative Property of Multiplication

$3 \times 6 = 6 \times 3$

Lesson 3–7

decompose

6

$6 \times 1 = 6$
$3 \times 2 = 6$
$2 \times 3 = 6$
$1 \times 6 = 6$

Factors: 1, 2, 3, 6

Lesson 3–1

dividend

$64 \div 8 = 8$

Lesson 3–1

divisor

$3\overline{)19}$

Lesson 3–1

fact family

$4 \times 7 = 28, 7 \times 4 = 28,$
$28 \div 7 = 4, 28 \div 4 = 7$

Lesson 3–1

factor

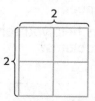

$2 \times 2 = 4$

Lesson 3–5

Identity Property of Multiplication

$1 \times 10 = 10 \qquad 10 \times 1 = 10$

Ideas for Use

The order in which two numbers are multiplied does not change the product.

How is this property different from the Associative Property of Multiplication?

_____

_____

The grouping of the factors does not change the product.

How is this property like the Associative Property of Addition?

_____

_____

A number that is being divided.

Write the words from this set of cards that are related to *dividend*.

_____

To break apart a number.

Decompose is a multiple-meaning word. Use another meaning of *decompose* in a sentence.

_____

_____

A group of related facts using the same numbers.

How can you remember that a fact family uses the same 3 numbers?

_____

_____

_____

The number by which the dividend is being divided.

The suffix *-or* means "a thing that does the action". Write how *divisor* does the action in division.

_____

_____

When any number is multiplied by 1, the product is that number.

Look in the dictionary for a meaning of *identity*. Write a sentence using that meaning.

_____

_____

A number that divides a whole number evenly. Also, a number that is multiplied by another number.

How can factors help you solve multiplication and division problems?

_____

_____

MY Vocabulary Cards

Mathematical
PRACTICE

multiple

multiples of 8:

0,   8,   16,   24,   32...

0×8 1×8 2×8 3×8 4×8

product

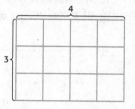

$3 \times 4 = 12$

quotient

$15 \div 3 = 5$

repeated subtraction

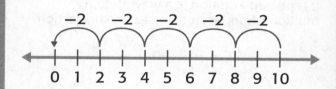

Zero Property of Multiplication

$12 \times 0 = 0$ $3 \times 0 = 0$

Ideas for Use

- Write a tally mark on each card every time you read or write the word. Challenge yourself to use at least 3 tally marks for each card.

- Use the blank cards to write your own vocabulary cards.

The answer to a multiplication problem.

Product is a multiple-meaning word. Use another meaning of *product* in a sentence.

_____

_____

A multiple of a number is the product of that number and a whole number.

Write a clue to help you remember that multiples are used in multiplication.

_____

_____

Subtraction of the same subtrahend over and over again.

If repeated addition is a way of doing multiplication, what is repeated subtraction?

_____

The answer to a division problem.

Write a division equation whose quotient is 4.

_____

When any number is multiplied by 0, the product is 0.

Write a tip to help you remember this property.

_____

_____

_____

MY Foldable

FOLDABLES® Follow the steps on the back to make your Foldable.

Multiples

| 4 | 5 | 6 | 12 | 15 | 24 | 27 |

Factors

| 4 | 5 | 6 | 12 | 15 | 24 | 27 |

FOLDABLES®
Study Organizer

Multiples

Factors

Name ....

Relate Multiplication and Division

Lesson 1

ESSENTIAL QUESTION
How are multiplication and division related?

You can use models to represent multiplication and division. Multiplication and division are opposite, or inverse, operations.

High score!

 Math in My World Tools Watch Tutor

Example 1

Felisa and Mei-Ling went to an arcade. They played 4 games that cost $3 each. What was the total cost?

Write related multiplication and division sentences to solve.

1 Arrange counters in an array with 3 rows and 4 columns. Draw the counters in the chart.

There are _____ counters total.

2 Write a multiplication sentence.

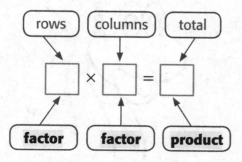

rows   columns   total

☐ × ☐ = ☐

factor   factor   product

3 Write a related division sentence.

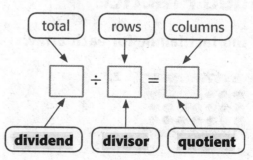

total   rows   columns

☐ ÷ ☐ = ☐

dividend   divisor   quotient

So, the total cost was $_____.

Online Content at connectED.mcgraw-hill.com

A **fact family** is a set of four related multiplication and division facts that use the same three numbers.

Example 2

Latanya and her father made an array of buttons. Write a fact family for the array.

There are 3 rows, 4 columns, and a total of 12 buttons.

☐ × ☐ = ☐   ☐ ÷ ☐ = ☐

☐ × ☐ = ☐   ☐ ÷ ☐ = ☐

Example 3

Vanessa has 36 books to put on 4 shelves. The same number of books will be placed on each shelf. How many books will be on each shelf?

Find _____ ÷ _____ . You can use a related multiplication fact to help you divide.

$36 ÷ 4 = \blacksquare$ ←— [Find the unknown.]

THINK: $4 × \blacksquare = 36$

$4 × \boxed{} = 36$

So, $36 ÷ 4 = \boxed{}$.

Vanessa will place $\boxed{}$ books on each shelf.

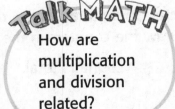

Talk MATH

How are multiplication and division related?

Guided Practice

Write the fact family for each array.

1.

2.

_____

_____

Independent Practice

Write the fact family for each array or set of numbers.

3.

_____

_____

_____

4.

_____

_____

_____

5. 6, 9, 54

_____

_____

_____

6. 7, 8, 56

_____

_____

_____

7. 9, 11, 99

_____

_____

_____

8. 11, 12, 132

_____

_____

_____

Find each unknown to complete each fact family.

9. $4 \times 8 =$ _____

_____ $\times 4 = 32$

$32 \div$ _____ $= 8$

$32 \div 8 =$ _____

10. _____ $\times 9 = 72$

$9 \times 8 =$ _____

$72 \div$ _____ $= 8$

$72 \div 8 =$ _____

Problem Solving

11. **Mathematical PRACTICE 4** **Model Math** Five banjo players entered the banjo contest. They are each playing four-string banjos. How many strings do the banjos have in all?

$5 \times 4 =$ _____

12. Ed wants to share 18 grapes equally among himself and two friends. How many grapes will each person receive?

$18 \div 3 =$ _____

13. A Seminole basket-maker made a small basket. He used 9 blades of sweet grass and braided them for each coil. He made 9 coils. How many blades of sweet grass did he use to make the basket?

$9 \times 9 =$ _____

My Work!

HOT Problems

14. **Mathematical PRACTICE 2** **Use Number Sense** Draw an array. Write a fact family for your array.

_____

_____

15. **?** **Building on the Essential Question** How can fact families and multiplication facts help you divide? Explain.

_____

_____

_____

MY Homework

Homework Helper

Need help? connectED.mcgraw-hill.com

Write a fact family for the array.

> There are 2 rows, with 4 in each row. That makes a total of 8. The numbers in the fact family are 2, 4, and 8.

$2 \times 4 = 8$ $8 \div 2 = 4$
$4 \times 2 = 8$ $8 \div 4 = 2$

Find $14 \div 2$. Use a related multiplication fact.

Think: $2 \times \blacksquare = 14$

$2 \times 7 = 14$

> Using the related multiplication fact $2 \times 7 = 14$, you know that $14 \div 2 = 7$.

So, $14 \div 2 = 7$.

Practice

Write a fact family for each set of numbers.

1. 3, 6, 18

2. 2, 5, 10

_____   _____

_____   _____

_____   _____

Problem Solving

3. Monique volunteered 24 hours last month at the animal shelter. If she volunteered the same amount of hours each week for 4 weeks, how many hours did she volunteer each week?

_____

4. **Mathematical PRACTICE** **Model Math** Tyler took 36 pictures while he was on vacation. He wants to put them into a photo album. He will put 6 pictures on each page. How many pages will he use?

_____

5. Lani's mother gave her and her two sisters $21 to spend at the movies. If each girl gets the same amount, how much do they each have to spend?

_____

Vocabulary Check

6. Use the vocabulary terms listed below to label each part of the equations.

dividend   divisor   factor   product   quotient

$$24 \div 4 = 6 \qquad 4 \times 6 = 24$$

_____ _____ _____   _____ _____ _____

Match each term with its definition.

7. division

8. fact family

9. multiplication

- a group of related facts using the same numbers

- an operation on two numbers to find the product

- an operation on two numbers to find the quotient

Test Practice

10. Which is a related multiplication fact for $18 \div \blacksquare = 6$?

Ⓐ $18 \div 2 = 9$ Ⓑ $6 \times 3 = 18$ Ⓒ $18 - 12 = 6$ Ⓓ $6 \times 4 = 24$

Need more practice? Download Extra Practice at **connectED.mcgraw-hill.com**

Relate Division and Subtraction

Lesson 2

ESSENTIAL QUESTION
How are multiplication and division related?

You know that repeated addition can be used to multiply. **Repeated subtraction** can be used to divide.

 ## Math in My World

 Watch Tutor

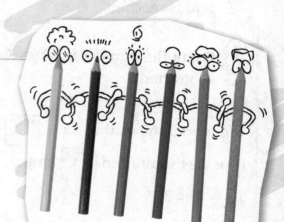

Example 1

Wyatt is giving 15 colored pencils to 3 friends. How many colored pencils will each friend get?

You can use repeated subtraction to find 15 ÷ 3.

1   2   3   4   5

$$\begin{array}{r} 15 \\ -\ 3 \\ \hline \square \end{array} \qquad \begin{array}{r} \square \\ -\ 3 \\ \hline \square \end{array} \qquad \begin{array}{r} \square \\ -\ 3 \\ \hline \square \end{array} \qquad \begin{array}{r} \square \\ -\ 3 \\ \hline \square \end{array} \qquad \begin{array}{r} \square \\ -\ 3 \\ \hline \square \end{array}$$

How many times was 3 subtracted from 15? _____

15 ÷ 3 = _____ .

So, each friend will get _____ colored pencils.

Example 2

The students in Mr. Bantha's class are helping set up games for Family Fun Math Night. Each game can have four players. How many games will be needed for 12 people?

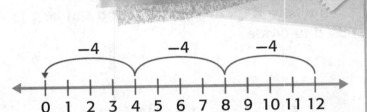

Find 12 ÷ 4.

You can skip count backwards on a number line to find 12 ÷ 4.

 Find **12**.

 Skip count backwards by **4** until zero is reached.

0 1 2 3 4 5 6 7 8 9 10 11 12

3 Count the **number of times** 4 was subtracted.

The model shows that **12** − ☐ − ☐ − ☐ = ☐.

Four was subtracted ☐ times.

12 ÷ 4 = 3

So, _____ games will be needed.

Talk MATH

Describe how to use subtraction to find 16 ÷ 4 without using a number line.

Guided Practice ✓

Use repeated subtraction to divide.

1. 10 ÷ 2 = ☐

10 − 2 = ☐

☐ − 2 = ☐

☐ − 2 = ☐

☐ − 2 = ☐

☐ − 2 = ☐

2. 12 ÷ 3 = ☐

12 − 3 = ☐

☐ − 3 = ☐

☐ − 3 = ☐

☐ − 3 = ☐

Independent Practice

Use repeated subtraction to divide.

3. 16 ÷ 8 = _____

4. 14 ÷ 2 = _____

5. 18 ÷ 6 = _____

6. 15 ÷ 5 = _____

7. 25 ÷ 5 = _____

8. 27 ÷ 9 = _____

9. 24 ÷ 8 = _____

10. 20 ÷ 4 = _____

11. 24 ÷ 6 = _____

Algebra Find each unknown number.

12. 12 ÷ 4 = ■

■ = _____

13. 21 ÷ ■ = 3

■ = _____

14. ■ ÷ 5 = 2

■ = _____

Problem Solving

Write a number sentence to solve.

15. Mike buys a DVD set of a television series. Each DVD has 6 episodes. There are 24 episodes. How many DVDs are in the set?

_____

16. Lucy is making the same bracelet for 8 of her friends. She has 32 beads. How many beads can she put on each bracelet?

_____

17. **Mathematical PRACTICE 8** **Look for a Pattern** There are 9 mockingbird nests in a tree. There are a total of 18 eggs in the nests. Each nest has the same number of eggs. How many mockingbird eggs are in each nest?

_____

HOT Problems

18. **Mathematical PRACTICE 3** **Find the Error** Olivia is using repeated subtraction to find $18 \div 2$. Find and correct her mistake.

$$18 - 2 - 2 - 2 - 2 - 2 - 2 - 2 - 2 = 2$$

_____

_____

19. **Building on the Essential Question** How are the operations of subtraction and division related? Explain.

_____

_____

_____

MY Homework

Lesson 2

Relate Division and Subtraction

Homework Helper

eHelp

Need help? connectED.mcgraw-hill.com

Find 12 ÷ 2. Use repeated subtraction.

$$\begin{array}{ccccc} 12 & 10 & 8 & 6 & 4 & 2 \\ -2 & -2 & -2 & -2 & -2 & -2 \\ \hline 10 & 8 & 6 & 4 & 2 & 0 \end{array}$$

2 was subtracted 6 times

So, 12 ÷ 2 = 6.

Practice

Use repeated subtraction to divide.

1. 27 ÷ 3 = _____

2. 30 ÷ 10 = _____

3. 24 ÷ 6 = _____

4. 15 ÷ 1 = _____

5. 14 ÷ 7 = _____

6. 18 ÷ 3 = _____

7. 10 ÷ 5 = _____

8. 28 ÷ 4 = _____

9. 20 ÷ 4 = _____

Problem Solving

Write a number sentence for each situation. Then solve.

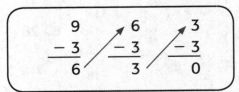

10. **Mathematical PRACTICE** 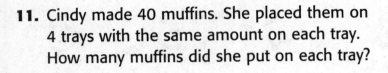 **Model Math** A bag of 15 apples was shared among 5 boys equally. How many apples did each boy get?

_____

11. Cindy made 40 muffins. She placed them on 4 trays with the same amount on each tray. How many muffins did she put on each tray?

_____

12. Shiloh put 24 eggs into 3 bowls with the same amount in each bowl. How many eggs did she put in each bowl?

_____

Vocabulary Check

13. Explain how you would use repeated subtraction to find $8 \div 2$.

_____

_____

_____

Test Practice

14. Which number sentence is represented by the repeated subtraction at the right?

$$\begin{array}{ccc} 9 & 6 & 3 \\ -3 & -3 & -3 \\ \hline 6 & 3 & 0 \end{array}$$

   (A) $9 \div 3 = 6$ (B) $6 \div 3 = 2$

   (C) $9 \div 9 = 1$ (D) $9 \div 3 = 3$

Name

Multiplication as Comparison

Lesson 3
ESSENTIAL QUESTION
How are multiplication
and division related?

Sometimes a problem uses a phrase like *times as many*, *times more*, and *times as much*. These kinds of problems are comparison problems.

 Math in My World Tools Watch Tutor

Example 1

Mary attended camp for 7 days this summer. Tyler attended 3 times as many days as Mary. Find the number of days Tyler attended camp.

Use counters to help you compare the groups of days.

1 Model Mary's days at camp as _____ group of 7 days. Draw your model.

2 Tyler attended 3 times as many days at camp.

Model Tyler's days at camp as _____ groups of 7. Draw your model.

My Drawing!

3 Find the total of 3 groups of 7.

_____ + _____ + _____ = _____

or

_____ × _____ = _____

So, Tyler attended camp for _____ days.

Online Content at connectED.mcgraw-hill.com

A type of model drawing is the bar diagram. A bar diagram can help you understand a problem and plan to solve it.

Example 2

Suki used 15 beads to make a bracelet. This is 3 times as many beads as what Cassady used. How many beads did Cassady use?

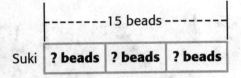

1 The bar diagram models this problem.

Cassady | ? beads |

|--------15 beads--------|

Suki | ? beads | ? beads | ? beads |

Suki used _____ times as many beads as Cassady.

2 Find how many beads Cassady used. Write an equation.

$3 \times ? = 15$ $3 \times$ _____ $= 15$ beads   ← Use a fact family.

So, Cassady used _____ beads.

Guided Practice

1. Use multiplication or division to complete the equation for the phrase below.

3 times as much

$3 \times 3 = ?$

$3 \times 3 =$ _____

Talk MATH

One way to interpret $24 = 8 \times 3$ is to say that 24 is 8 times as many as 3. What is another way you can interpret this equation?

Independent Practice

Use multiplication or division to complete each equation and/or drawing.

2. 3 times as many

_____ × _____ = 3

3. 5 times more

_____ × _____ = 25

4. 4 times as much

_____ × 3 = 12

5. 10 times as much

10 × _____ = 40

6. 2 times more

2 × _____ = 6

7. twice as many

2 × _____ = 14

Complete each bar diagram. Then complete the multiplication equation.

8. twice as many
as 4 boys

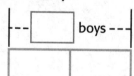

2 × _____ = 8

9. 2 times as many
as 3 bows

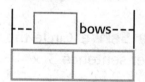

_____ × _____ = _____

10. 4 times as many
as 6 fish

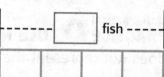

_____ × _____ = _____

Draw a bar diagram. Then write a multiplication equation.

11. 3 times as much money as $6

Write the equation.

_____

12. 5 times as many as 1 star

Write the equation.

_____

Problem Solving

Woof woof!

Draw a bar diagram and write an equation to solve.

13. **Mathematical PRACTICE 4 Model Math** There are 3 times as many blue balloons as green balloons. There are 4 green balloons. How many blue balloons are there?

My Work!

_____

Find the unknown.

_____ is 3 times as many as 4.

14. Nan needs 4 times as much flour as sugar. She needs 4 cups of sugar. How much flour does she need?

_____

Find the unknown.

_____ is 4 times as many as 4.

HOT Problems

15. **Mathematical PRACTICE 2 Use Number Sense** Circle the example that does not represent the number sentence $3 \times 4 = 12$. Explain.

$$4 + 4 + 4 = 12$$

$$12 - 4 = 8$$

_____

_____

16. **Building on the Essential Question** How can a bar diagram help me plan and solve a problem? Explain.

_____

_____

Name

MY Homework

Homework Helper

Need help? connectED.mcgraw-hill.com

Write a multiplication equation to describe the model.

5 times more

You need to find the total of 5 groups of 6.

Write a multiplication equation.

$6 \times 5 = 30$

You can also show this in a bar diagram.

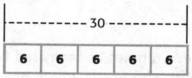

Practice

Write a multiplication equation to describe each model.

1. 4 times more

_____

2. 2 times as much

_____

3. 6 times as many

_____

4. twice as many

_____

5. 3 times more

_____

6. 5 times as much

_____

7. 4 times as much money as $5

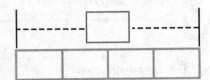

4 × _____ = _____

8. 2 times as many books as 7 books

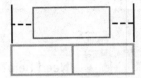

_____ × _____ = _____

Problem Solving

9. Henry has 3 gerbils. Hannah has 3 times as many. How many gerbils does Hannah have?

_____

10. Dani needs 6 times as many red beads as gold beads. She needs 7 gold beads. How many red beads does Dani need?

_____

11. Draw a bar diagram to represent 6 times more than $4.

Test Practice

12. Amelia swam four times as many laps as Wesley. Wesley swam 7 laps. How many laps did Amelia swim?

Ⓐ 11 laps   Ⓒ 35 laps

Ⓑ 28 laps   Ⓓ 21 laps

Name ....

Compare to Solve Problems

Additive comparisons use addition or subtraction to compare. Multiplicative comparisons use multiplication or division to compare.

Additive Comparison	Multiplicative Comparison
how much more	how many times more
how many more	how many times greater
how much less	

Math in My World

  Tutor

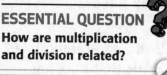
Surf's up!

Example 1

Bryan has been to a water park 4 times. Sarah has been to the water park three times as many times as Bryan. How many times has Sarah been to the water park?

Write an equation to find the unknown. You can use a letter, or a variable, to represent the unknown.

$4 \times \underline{\hspace{1cm}} = s$ ⟵ the unknown: the number of times Sarah has been to the water park

Draw a picture to show 3 times as many as 4, or 3 groups of 4.

The picture shows a total of 12.

So, $4 \times 3 = 12$.

Since $s = \underline{\hspace{1cm}}$, Sarah has been to the

water park $\underline{\hspace{1cm}}$ times.

My Drawing!

Online Content at ↗ **connectED.mcgraw-hill.com**

Example 2

Tutor

Jess has 18 baseball cards. She has 6 times as many baseball cards as Max. How many baseball cards does Max have?

Write an equation. Let ■ represent the number of baseball cards that Max has.

■ × _____ = _____

Since ■ × 6 = 18, you know that 18 ÷ 6 = ■. ◄— | Use a fact family.

Draw 6 equal groups of squares. Draw a total of 18 squares.

My Drawing!

There are _____ squares in each group.

■ = 3. So, Max has _____ baseball cards.

Talk MATH

How can unknown numbers be represented in equations?

Guided Practice ✓ Check

1. **Sue has 3 times as many beads as Jo. Jo has 8 beads. How many beads does Sue have? Write an equation to find the unknown number. Use a variable for the unknown.**

   3 × _____ = b

   b = _____

Independent Practice

Algebra Write an equation to find the unknown number. Use a symbol for the unknown.

2. Paul drew 4 times as many pictures as Dennis. Paul drew 16 pictures. How many pictures did Dennis draw?

_____

3. Maria made 21 cupcakes. This is three times as many as the number of cupcakes that Mary made. How many cupcakes did Mary make?

_____

Algebra Write an equation to find the unknown number. Use a variable for the unknown.

4. Wendy has dance class 2 days a week. James has dance class 5 times a week. How many more times a week does James have dance class than Wendy?

_____

5. There are 4 fewer white bunnies than gray bunnies. There are 9 white bunnies. How many gray bunnies are there?

_____

Use the table for Exercises 6–9.

6. How many more shoes were sold than belts?

_____

7. Which item sold 2 times as many as shirts?

_____

8. Which exercises on this page used addition or subtraction to compare? List them.

_____

9. Which exercises on this page used multiplication or division to compare? List them.

_____

Items Sold at a Department Store	
Item	**Number Sold**
hats	4
shoes	7
belts	2
shirts	8
pants	16
socks	12

Problem Solving

10. Jerry read 24 pages this weekend. That is 4 times as many pages as what he read last weekend. How many pages did he read last weekend?

_____

11. A bean plant is 3 inches tall. A corn plant is five times as tall. How many inches tall is the corn plant?

_____

12. **Mathematical PRACTICE ➊ Plan Your Solution** Hannah used 10 more blocks than Steve. Steve used 7 blocks. How many blocks did Hannah use?

_____

13. There are 10 fewer robins than cardinals. There are 16 cardinals. How many robins are there?

_____

HOT Problems

14. **Mathematical PRACTICE ➊ Check for Reasonableness** Missy earned three times as many points as Jerry. Kimmie earned 9 more points than Jerry. Missy earned 21 points. How many points did Kimmie earn?

_____

15. **? Building on the Essential Question** How can you tell the difference between additive comparison and multiplicative comparison?

_____

_____

Name ....

MY Homework

Homework Helper

Need help? connectED.mcgraw-hill.com

Additive comparisons use addition or subtraction to compare.

Multiplicative comparisons use multiplication or division to compare.

Maya went swimming 7 times this month. Her brother went swimming 14 times this month. How many times more did Maya's brother go swimming than Maya?

Write an equation. Let b represent the unknown.

$7 \times b = 14$

$7 \times 2 = 14$

So, $b = 2$.

Maya's brother went swimming 2 times as much as Maya.

Additive Comparison
how much more
how many more
how much less

Multiplicative Comparison
how many times more
how many times greater

Practice

Algebra **Write an equation to find the unknown number. Use a symbol for the unknown.**

1. Julie earned $25. This is 5 times as much as what Lisa earned. How much did Lisa earn?

_____

2. The red team scored 4 goals. The blue team scored 3 times as many goals. How many goals did the blue team score?

_____

Problem Solving

Write an equation to find the unknown number.
Use a variable for the unknown.

3. A large aquarium has 6 more fish than a small aquarium. There are 19 fish in a large aquarium. How many fish are in a small aquarium?

_____

4. **Mathematical** **PRACTICE** **1** **Plan Your Solution** The table shows the number of roller coasters each student rode at the amusement park.

Riding Roller Coasters	
Student	Number of Roller Coasters
Sarah	18
Tom	15
Val	3
Warren	9

How many more roller coasters has Tom been on than Val?

_____

Who rode a roller coaster twice as many times as Warren?

_____

Test Practice

5. Which shows 7 times as many as 5?

Ⓐ 2

Ⓑ 5

Ⓒ 12

Ⓓ 35

Vocabulary Check

1. Write each word from the word bank on the correct label.

dividend   **divisor**   **fact family**

factor   **product**   **quotient**

$3 \times 2 = 6$

$2 \times 3 = 6$

$6 \div 3 = 2$

$6 \div 2 = 3$

Concept Check

Use repeated subtraction to divide.

2. $18 \div 6 = $ _____

3. $28 \div 7 = $ _____

Write the fact family for each set of numbers.

4. 6, 4, 24

5. 7, 6, 42

6. 8, 4, 32

_____

_____

_____

_____

7. Jerry made lemonade with the lemons shown.

Write a fact family for the array of lemons.

_____

8. There are 21 fish in an aquarium at the zoo. In order to clean the aquarium, the zookeeper must put the fish in three smaller tanks. If the same number of fish are in each tank, how many fish are in each tank? Use repeated subtraction to solve.

_____

My Work!

9. Algebra Gina has 5 pencils. Her sister has twice as many pencils. Write an equation to find the unknown number. Use a variable for the unknown.

_____

Test Practice

10. Find the missing number. $15 \div 3 = \blacksquare$.

   Ⓐ 1   Ⓒ 5

   Ⓑ 3   Ⓓ 12

Multiplication Properties and Division Rules

Lesson 5

ESSENTIAL QUESTION
How are multiplication and division related?

 ## Math in My World [Tutor]

Jenny	Cliff
Pack lunches	Set table
Take out trash	Clean room
	Walk dog

Example 1

The table shows Jenny's and Cliff's chores. Jenny earns $3 for each chore and Cliff earns $2 for each chore. How much does each person earn for completing chores?

Complete each number sentence.

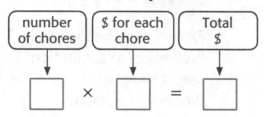

Jenny

number of chores	$ for each chore	Total $

☐ × ☐ = ☐

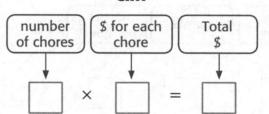

Cliff

number of chores	$ for each chore	Total $

☐ × ☐ = ☐

So, each person earns _____ . The order in which the factors are multiplied does not change the product.

Key Concept Multiplication Properties

Commutative Property of Multiplication
When multiplying, the order of the factors does not change the product.

$4 \times 2 = 8$
$2 \times 4 = 8$

Identity Property of Multiplication
When any number is multiplied by 1, the product is that number.

$4 \times 1 = 4$

Zero Property of Multiplication
When any number is multiplied by 0, the product is 0.

$3 \times 0 = 0$

Helpful Hint

You will learn about the Associative Property of Multiplication and the Distributive Property in later lessons.

The following rules can help you with division.

Key Concept Division Rules

Zeros in Division
When you divide 0 by any nonzero number, the quotient is 0.

$0 \div 5 = 0$

It is not possible to divide a number by 0.

Ones in Division
When you divide any number by 1, the quotient is always the dividend.

$7 \div 1 = 7$

When you divide any nonzero number by itself, the quotient is always 1.

$4 \div 4 = 1$

Helpful Hint

Quotient is the answer to a division problem.

Dividend is the number being divided.

Example 2

  Watch Tutor

There are 9 party favors and 9 guests. How many party favors will each guest get?

Complete the number sentence.

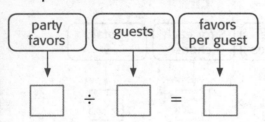

party favors ▼   guests ▼   favors per guest ▼

☐ ÷ ☐ = ☐

A nonzero number divided by the same number is 1.

So, each guest will get _____ party favor.

Talk MATH

Explain why the Identity Property of Multiplication uses 1 while the Identity Property of Addition uses 0.

Guided Practice Check ✓

Identify the property or rule shown by each equation.

1. $12 \times 0 = 0$

2. $8 \times 5 = 5 \times 8$

_____

_____

Independent Practice

Identify the property or rule shown by each equation.

3. $6 \div 1 = 6$

_____

_____

4. $10 \div 10 = 1$

_____

_____

5. $8 \times 0 = 0$

_____

_____

6. $0 \div 12 = 0$

_____

_____

7. $22 \times 1 = 22$

_____

_____

8. $4 \times 3 = 3 \times 4$

_____

_____

Algebra Find each unknown number. Identify the property or rule.

9. $3 \div \blacksquare = 1$

$\blacksquare =$ _____

_____

_____

10. $\blacksquare \times 8 = 8 \times 4$

$\blacksquare =$ _____

_____

_____

11. $\blacksquare \div 11 = 0$

$\blacksquare =$ _____

_____

_____

12. $\blacksquare \times 1 = 15$

$\blacksquare =$ _____

_____

_____

Problem Solving

13. On a hiking trip, Tamika and Bryan hiked 7 miles a day. They hiked for 5 days. Kurt and Sade hiked 5 miles a day. How many days did it take Kurt and Sade to hike the same distance as Tamika and Bryan? Write a number sentence to solve.

_____

14. There are 6 shelves in the library that each have 8 books displayed. How many books are on all of the shelves? Use the Commutative Property to write the multiplication sentence in two ways. Then solve.

_____

15. Mathematical **PRACTICE** **7** **Identify Structure** Explain why it is helpful to understand the Identity Property of Multiplication.

_____

_____

_____

HOT Problems

16. Mathematical **PRACTICE** **1** **Make a Plan** Write a multiplication problem that uses the Commutative Property of Multiplication to solve.

_____

_____

17. **?** **Building on the Essential Question** How do multiplication properties and division rules help you to multiply and divide?

_____

_____

Number and Operations in Base Ten
4.NBT.5

CCSS

MY Homework

Homework Helper

Need help? connectED.mcgraw-hill.com

The tables show the properties of multiplication and the division rules that can be used to help solve problems. Identify the property or rule used in the equation $5 \times 1 = 5$.

Multiplication Properties	
Commutative Property of Multiplication When multiplying, the order of the factors does not change the product.	$3 \times 4 = 12$ $4 \times 3 = 12$
Identity Property of Multiplication When any number is multiplied by 1, the product is that number.	$7 \times 1 = 7$
Zero Property of Multiplication When any number is multiplied by 0, the product is 0.	$6 \times 0 = 0$

Division Rules	
Zeros in Division When you divide 0 by any nonzero number, the quotient is 0. It is not possible to divide a number by 0.	$0 \div 9 = 0$
Ones in Division When you divide any number by 1, the quotient is always the dividend. When you divide any nonzero number by itself, the quotient is always 1.	$8 \div 1 = 8$ $6 \div 6 = 1$

The equation $5 \times 1 = 5$ shows the Identity Property of Multiplication.

Practice

Identify the property or rule shown by each equation.

1. $9 \div 1 = 9$

2. $33 \times 1 = 33$

Problem Solving

Complete each number sentence. Identify the property or rule.

3. $5 \div$ _____ $= 5$

_____

4. $9 \times 8 = 8 \times$ _____

_____

5. _____ $\div 12 = 0$

_____

6. **Mathematical PRACTICE 7** **Identify Structure** Dennis has 3 packs of pens with 2 pens in each pack. He has 2 packs of pencils with 3 pencils in each pack. Write two multiplication sentences to show how many pens and pencils he has.

_____

Vocabulary Check

Write a number sentence for each rule or property.

7. Ones in Division _____

8. Commutative Property of Multiplication _____

9. Zeros in Division _____

10. Zero Property of Multiplication _____

11. Identity Property of Multiplication _____

Test Practice

12. The Zero Property of Multiplication tells you that 25×0 is equal to what number?

Ⓐ 0   Ⓒ 7

Ⓑ 1   Ⓓ 25

Number and Operations in Base Ten
4.NBT.5

CCSS

The Associative Property of Multiplication

Lesson 6
ESSENTIAL QUESTION
How are multiplication and division related?

The **Associative Property of Multiplication** shows that the way in which numbers are grouped does not change their product.

 ## Math in My World Watch ▶ Tutor 💬

Example 1

There are 2 video games in each value pack. There are 6 value packs in each box. If Raul buys 3 boxes for his collection, how many video games will he have?

You need to find $2 \times 6 \times 3$. There are two ways to group the numbers.

One Way

Multiply 2×6 first.

$2 \times 6 \times 3 = (2 \times 6) \times 3$

$$= \boxed{} \times \boxed{}$$

Helpful Hint

Use repeated addition to find 12×3.

$12 + 12 + 12 = \boxed{}$

$$= \boxed{}$$

Another Way

Multiply 6×3 first.

$2 \times 6 \times 3 = 2 \times (6 \times 3)$

$$= \boxed{} \times \boxed{}$$

Helpful Hint

Use repeated addition to find 18×2.

$18 + 18 = \boxed{}$

$$= \boxed{}$$

So, Raul will have _____ video games.

Example 2

Use the Associative Property of Multiplication to find 9 × 2 × 4.

Find 9 × 2 first.

$9 × 2 × 4 = (9 × 2) × 4$

$= \boxed{} × 4$

$= 18 + 18 + 18 + 18$

$= \boxed{}$

It is easier to find 9 × 8 than 18 × 4.

Find 2 × 4 first.

$9 × 2 × 4 = 9 × (2 × 4)$

$= 9 × \boxed{}$

$= \boxed{}$

Helpful Hint

Parentheses ( ) tell you which numbers to multiply first.

Guided Practice

Multiply. Use the Associative Property.

1. $5 × 3 × 3 = 5 × (3 × 3)$

$= \boxed{} × \boxed{}$

$= \boxed{}$

2. $4 × 2 × 7 = (4 × 2) × 7$

$= \boxed{} × \boxed{}$

$= \boxed{}$

3. $3 × 1 × 6 = (3 × 1) × 6$

$= \boxed{} × \boxed{}$

$= \boxed{}$

Talk MATH

Identify the order that makes it easier to multiply the factors in 9 × 4 × 2. Explain.

Independent Practice

Multiply. Use the Associative Property.

4. $6 \times 1 \times 5 = $ _____

5. $2 \times 2 \times 7 = $ _____

6. $7 \times 5 \times 2 = $ _____

7. $10 \times 2 \times 5 = $ _____

8. $9 \times 3 \times 3 = $ _____

9. $6 \times 2 \times 2 = $ _____

10. $2 \times 3 \times 7 = $ _____

11. $9 \times 2 \times 4 = $ _____

12. $5 \times 1 \times 10 = $ _____

Compare. Use >, <, or =.

13. $4 \times 2 \times 9$ $\bigcirc$ $7 \times 4 \times 2$

14. $6 \times 2 \times 6$ $\bigcirc$ $5 \times 2 \times 8$

Find the value of each number sentence if ✹ = 2, ☺ = 3, and ★ = 4.

15. $5 \times 1 \times $ ★ $= $ _____

16. $6 \times $ ✹ $\times 3 = $ _____

17. ☺ $\times 3 \times $ ★ $= $ _____

Algebra Find the unknown number.

18. $4 \times $ ■ $\times 1 = 12$

■ $= $ _____

19. $2 \times 5 \times $ ■ $= 60$

■ $= $ _____

20. ■ $\times 3 \times 4 = 24$

■ $= $ _____

Problem Solving

Write a number sentence to solve.

21. Gabriel is training for a race. On days that he jogs, he jogs 2 miles. His exercise schedule is shown. How many miles will he jog in 6 weeks?

Exercise Schedule	
Day	**Activity**
Monday	jog
Tuesday	jog
Wednesday	basketball
Thursday	jog
Friday	basketball
Saturday	jog
Sunday	rest

_____

_____

22. **Mathematical PRACTICE 4 Model Math** Bianca bikes 2 miles to her grandfather's house and 2 miles back to her house 5 times each month. How many miles does she bike each month?

_____

_____

HOT Problems

23. **Mathematical PRACTICE 2 Reason** Circle the equation that does not belong with the other three. Explain.

| $4 \times \blacksquare \times 7 = 56$ | $5 \times 2 \times \blacksquare = 40$ | $\blacksquare \times 3 \times 9 = 54$ | $4 \times \blacksquare \times 5 = 40$ |

_____

_____

24. **? Building on the Essential Question** How does the Associative Property of Multiplication help you to calculate products mentally?

_____

_____

_____

MY Homework

Homework Helper

Need help? connectED.mcgraw-hill.com

Find **3 × 4 × 2.**

One Way

Multiply 3 × 4 first.

$$3 \times 4 \times 2 = (3 \times 4) \times 2$$
$$= \quad 12 \quad \times \quad 2$$
$$= \quad\quad 24$$

So, 3 × 4 × 2 = 24.

Another Way

Multiply 4 × 2 first.

$$3 \times 4 \times 2 = 3 \times (4 \times 2)$$
$$= 3 \times \quad 8$$
$$= \quad\quad 24$$

Practice

Multiply. Use the Associative Property.

1. 5 × 2 × 7 = _____

2. 8 × 3 × 2 = _____

3. 4 × 2 × 5 = _____

4. 5 × 4 × 3 = _____

5. 8 × 2 × 2 = _____

6. 3 × 2 × 5 = _____

Algebra **Find the unknown in each equation.**

7. 4 × _____ × 8 = 64

8. 3 × 4 × _____ = 120

9. 4 × 2 × _____ = 40

10. 6 × 2 × _____ = 96

Problem Solving

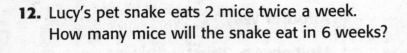

Mathematical PRACTICE 2 **Use Algebra** Write an equation to solve.

11. A bus seats 4 people in a row, and there are 12 rows. How many people would fit on 2 buses?

_____

12. Lucy's pet snake eats 2 mice twice a week. How many mice will the snake eat in 6 weeks?

_____

13. Tyler delivers newspapers. He collected $10 from each customer. He has 5 customers on each street, and he delivers to 4 different streets. How much money did Tyler collect?

_____

Vocabulary Check

Draw a line to match each property with the equation that represents it.

14. Associative Property of Multiplication

15. Commutative Property of Multiplication

• $8 \times 9 = 9 \times 8$

• $(4 \times 7) \times 2 = 4 \times (7 \times 2)$

Test Practice

16. Emerson runs 2 miles to her friend's house and 2 miles back to her house 5 times each month. How many miles does she run?

ⓐ 9 miles   ⓒ 20 miles

ⓑ 10 miles   ⓓ 50 miles

Number and Operations in Base Ten
4.OA.4

CCSS

Factors and Multiples

 Math in My World Tools Watch Tutor

Example 1

Mrs. Navarro is arranging desks in her classroom. There are 12 desks. How many ways can she arrange the desks so that the number of desks in each row is the same?

To find the different arrangements of desks, break down or **decompose** 12 into its factors. Think of factors that result in a product of 12. Use factors to write a number sentence for the arrays shown.

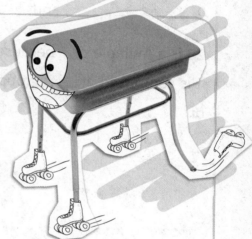

□ × □ = □

□ × □ = □

□ × □ = □

Helpful Hint
There are three more possible arrays:

12 × □

6 × □

4 × □

The factors of 12 are ____, ____, ____, ____, ____, and ____ .

So, the desks can be arranged in ____ ways.

A **multiple** of a number is the product of that number and any whole number. For example, 15 is a multiple of 5 because it is composed, or made, of 3 groups of 5. 15 is also a multiple of 3.

Example 2

All of the numbers listed in Row 7 or Column 7 are multiples of 7. Shade the multiples of 7 on the chart.

The first five multiples of 7 are

0, _____, _____, _____, and _____.

28 is a multiple of 7 because it is composed of _____ groups of _____.

×	0	1	2	3	4	5	6	7	8	9	10
0	0	0	0	0	0	0	0	0	0	0	0
1	0	1	2	3	4	5	6	7	8	9	10
2	0	2	4	6	8	10	12	14	16	18	20
3	0	3	6	9	12	15	18	21	24	27	30
4	0	4	8	12	16	20	24	28	32	36	40
5	0	5	10	15	20	25	30	35	40	45	50
6	0	6	12	18	24	30	36	42	48	54	60
7	0	7	14	21	28	35	42	49	56	63	70
8	0	8	16	24	32	40	48	56	64	72	80
9	0	9	18	27	36	45	54	63	72	81	90
10	0	10	20	30	40	50	60	70	80	90	100

Talk MATH

Explain how factors and multiples are related.

Guided Practice

Find the factors of each number.

1. 6

2. 36

_____

_____

Identify the first five multiples.

3. 4

4. 9

_____

Independent Practice

Find the factors of each number.

5. 4

6. 7

7. 14

8. 28

9. 30

10. 35

List the first five multiples.

11. 1

12. 3

13. 5

14. 7

15. 8

16. 6

Tell the total number modeled by each array. Then find the factors of that number.

17.

18.

Problem Solving

19. Complete the Venn Diagram.

factors of 28   factors of both 20 and 28   factors of 20

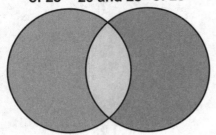

Time for a walk?

My Work!

20. Pedro walks his dog 3 times a day. How many times does Pedro walk his dog in one week? Find multiples of 3 to tell how many times Pedro walks his dog in 8, 9, and 10 days.

21. Mathematical **PRACTICE** 5 **Use Math Tools** There are 16 cans of soup on a shelf. One way the cans can be displayed is in a 1 × 16 array. Think of factors of 16 to identify two more ways the cans can be displayed.

HOT Problems

22. Mathematical **PRACTICE** 1 **Make a Plan** Identify the two numbers less than 20 with the most factors.

23. ❓ **Building on the Essential Question** How do you know when you have found all the factors of a number?

Number and Operations in Base Ten
4.OA.4

CCSS

MY Homework

Homework Helper

Need help? connectED.mcgraw-hill.com

Find the factors of 16.

Think of factors that result in a product of 16.

$1 \times 16 = 16$
$2 \times 8 = 16$
$4 \times 4 = 16$

So, the factors of 16 are 1, 2, 4, 8, and 16.

Identify the first five multiples of 4.

First multiple: $0 \times 4 = 0$
Second multiple: $1 \times 4 = 4$
Third multiple: $2 \times 4 = 8$
Fourth multiple: $3 \times 4 = 12$
Fifth multiple: $4 \times 4 = 16$

So, the first five multiples of 4 are 0, 4, 8, 12, and 16.

Practice

Find the factors of each number.

1. 14

2. 20

Identify the first five multiples for each number.

3. 2

4. 3

5. 6

6. 5

7. 8

8. 7

![Real World] **Problem Solving**

9. Each music class sings 8 songs a week. How many songs does each class sing in 5 weeks? 6 weeks? 7 weeks?

_____

10. **Mathematical PRACTICE 5** ▷ **Use Math Tools** Tyra wants to arrange 32 tiles in equal rows and columns. How many ways could she organize the tiles? List the factors.

_____

_____

11. Members of the marching band line up in 6 rows of 8. How many total members are there? Identify two other ways they could line up in equal rows and columns.

_____

_____

Vocabulary Check [Vocab abc]

Complete each sentence with the correct vocabulary term.

decompose   multiple

12. The number 12 is a _____ of 2, 3, and 4.

13. One way to _____ 24 is to show it as 2 × 12.

Test Practice

14. Which set of numbers correctly shows all of the factors of 28?

Ⓐ 1, 2, 4, 7, 14, 28   Ⓒ 1, 2, 7, 14, 28

Ⓑ 0, 1, 7, 14, 28   Ⓓ 1, 2, 4, 7, 8, 14, 28

Problem-Solving Investigation

STRATEGY: Reasonable Answers

Learn the Strategy

Starr won 4 tickets at Family Game Night. Suzi won 5 times as many tickets. Is it reasonable to say that they won 24 tickets altogether?

1 Understand

What facts do you know?

Starr won _____ tickets.

Suzi won _____ times as many tickets as Starr.

What do you need to find?

the total number of _____ won altogether

2 Plan

Find 5 × 4. Then add that to 4.

3 Solve

Find 5 × 4.

Model 5 groups of 4.

So, 5 × 4 = 20. Suzi won 20 tickets. Add. 20 + 4 = _____

Compare. The girls won _____ tickets. The answer is reasonable.

4 Check

Does my answer make sense? Explain.

_____

Online Content at connectED.mcgraw-hill.com

Practice the Strategy

Dasan delivers 283 newspapers each week. Lisa delivers 302 newspapers each week. Is 400 a reasonable estimate for the number of newspapers they deliver each week altogether?

 Understand

What facts do you know?

_____

_____

What do you need to find?

_____

_____

 Plan

_____

_____

_____

3 Solve

4 Check

Does my answer make sense? Explain.

_____

_____

Apply the Strategy

Determine a reasonable answer for each problem.

1. The table shows the number of pennies collected by four children. Is it reasonable to say that Myron and Teresa collected about 100 pennies in all? Explain.

Pennies Collected	
Child	**Number of Pennies**
Myron	48
Teresa	52
Veronica	47
Warren	53

Mathematical
2. **PRACTICE** 1 **Make Sense of Problems** Jay will make $240 doing yard work for 6 weeks. He is saving his money to buy camping equipment that costs $400. He has already saved $120. Is it reasonable to say that Jay will save enough money in 6 weeks? Explain.

_____

_____

_____

_____

3. Write a problem in which $180 would be a reasonable answer.

_____

_____

_____

Review the Strategies

4. A truck holds the number of cars shown. A parking lot has 6 times as many cars. How many cars are in the parking lot?

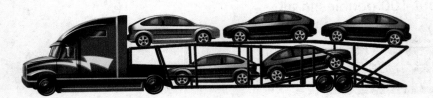

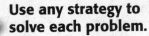

_____

5. **Mathematical PRACTICE 1** **Check for Reasonableness** Jack's basketball games are 4 quarters that are each 8 minutes long. Is it possible for Jack to play 35 minutes in a game? Explain.

_____

6. **Mathematical PRACTICE 3** **Find the Error** Mark and his dad are going to the amusement park. One roller coaster has 1,204 feet of track. Another roller coaster has 2,941 feet of track. Mark estimates that the two roller coasters have a total of about 3,000 feet of track. Find and correct his mistake.

_____

_____

_____

7. It costs $12 for 2 admissions to miniature golf. Abby wants to invite 9 of her friends. At this rate, how much would it cost for 10 people?

_____

MY Homework

Homework Helper

Need help? connectED.mcgraw-hill.com

Callie and Cole each used 24 craft sticks for their art project. Dinah and Dennis each used 33 craft sticks for their project. Is it reasonable to say they used about 100 craft sticks in all?

1 Understand

What facts do you know?

Callie and Cole each used 24 craft sticks.

Dinah and Dennis each used 33 craft sticks.

What do you need to find?

I need to determine if they used about 100 craft sticks in all.

2 Plan

I will round the number of craft sticks each pair of students used. Then I will add the rounded numbers.

3 Solve

Round 24 to the nearest ten: 20   Round 33 to the nearest ten: 30

Add: 20 + 20 + 30 + 30 = 100

So, it is reasonable to say the children used about 100 craft sticks in all.

4 Check

I will add the numbers.

24	33	48
+ 24	+ 33	+ 66
48	66	114

114 is close to 100, so the estimate is reasonable.

Problem Solving

Determine a reasonable answer for each problem.

1. Kevin can carry a basket 5 feet. Rachel can carry it 3 feet farther than Kevin. Daniel can carry the basket half as far as Rachel. Is it reasonable to say that they can carry it 15 feet with each person taking only one turn?

_____

_____

2. Josh and Anthony have a lemonade stand. They charge $1 for 2 cups of lemonade. They sell 14 cups each afternoon. Is it reasonable to say Josh and Anthony earned more than $50 after 3 afternoons?

_____

_____

3. Javier's grandmother lives 120 miles away. It takes 1 hour to travel 40 miles by train. If Javier leaves at 7 A.M., is it reasonable to say he will arrive in his grandmother's city by 9 A.M.?

_____

_____

4. **Mathematical PRACTICE 1** Make Sense of Problems
Brittany wants to make cookies for the whole fourth grade. One batch of dough makes 2 dozen cookies. There are 68 fourth graders at her school. Is it reasonable to say that Brittany will need more than two batches of dough?

_____

_____

_____

Vocabulary Check *Vocab abc*

Write the letter of each definition on the line next to the correct vocabulary word.

1. **Associative Property of Multiplication** _____

2. **Commutative Property of Multiplication** _____

3. **decompose** _____

4. **dividend** _____

5. **divisor** _____

6. **fact family** _____

7. **factor** _____

8. **Identity Property of Multiplication** _____

9. **multiple** _____

10. **product** _____

11. **quotient** _____

12. **repeated subtraction** _____

13. **Zero Property of Multiplication** _____

A. A group of related facts using the same numbers.

B. The number by which the dividend is being divided.

C. The property that states that the order in which two numbers are multiplied does not change the product.

D. The property that states any number multiplied by zero is zero.

E. A number that is multiplied by another number.

F. The answer of a division problem.

G. The answer of a multiplication problem.

H. The product of a given number and any whole number.

I. The property that states that the grouping of the factors does not change the product.

J. A strategy that can be used to divide.

K. A number that is being divided.

L. The property that states when any number is multiplied by 1, the product is that number.

M. A way to break down a number into its factors.

Concept Check

Write a fact family for each set of numbers.

14. 3, 7, 21 _____ _____ _____ _____

15. 9, 5, 45 _____ _____ _____ _____

Use repeated subtraction to divide.

16. 42 ÷ 7 = _____   **17.** 56 ÷ 8 = _____   **18.** 36 ÷ 9 = _____

---

19. Use multiplication to complete the number sentence.

5 times as many

     □ × □ = □

Identify the property or rule shown by each equation.

20. $6 \times 8 = 8 \times 6$ **21.** $(3 \times 2) \times 6 = 3 \times (2 \times 6)$

_____   _____

_____   _____

Find the factors of each number.

22. 16   **23.** 18   **24.** 15

_____   _____   _____

List the first five multiples.

25. 2   **26.** 10   **27.** 12

_____   _____   _____

Problem Solving

28. There are 8 cans of soup. There are 3 times as many cans of vegetables. How many cans of vegetables are there? Write an equation to find the unknown. Use a variable for the unknown.

_____

29. Claire has 15 green beads, 8 blue beads, and 4 yellow beads. If she puts them onto 3 strings equally, how many beads are on each string?

_____

30. Stefanie and Eva want to share the shells that they collected on their trip to the beach. They have 18 shells in all. Use related facts and draw an array that will help them decide how they can evenly divide their shells.

They can each get _____ shells.

31. If 7 people each ride a roller coaster 5 times, and it is $2 per person per ride, what is the total price they paid for the rides?

_____

Test Practice

32. Marina has scored 9 points on each of 11 quizzes. How many points has she scored in all?

Ⓐ 9 points   Ⓒ 90 points

Ⓑ 20 points   Ⓓ 99 points

Use what you learned about multiplication and
division to complete the graphic organizer.

Same

ESSENTIAL
QUESTION

How are multiplication
and division related?

Different

Now reflect on the ESSENTIAL QUESTION **Write your answer below.**

_____

_____

_____

ESSENTIAL QUESTION

How can I communicate multiplication?

Let's Go Shopping!

Watch

Watch a video!

MY Common Core State Standards

Number and Operations in Base Ten

4.NBT.1 Recognize that in a multi-digit whole number, a digit in one place represents ten times what it represents in the place to its right.

4.NBT.3 Use place value understanding to round multi-digit whole numbers to any place.

4.NBT.5 Multiply a whole number of up to four digits by a one-digit whole number, and multiply two two-digit numbers, using strategies based on place value and the properties of operations. Illustrate and explain the calculation by using equations, rectangular arrays, and/or area models.

Operations and Algebraic Thinking *This chapter also addresses these standards:*

4.OA.3 Solve multistep word problems posed with whole numbers and having whole-number answers using the four operations, including problems in which remainders must be interpreted. Represent these problems using equations with a letter standing for the unknown quantity. Assess the reasonableness of answers using mental computation and estimation strategies including rounding.

4.OA.4 Find all factor pairs for a whole number in the range of 1–100. Recognize that a whole number is a multiple of each of its factors. Determine whether a given whole number in the range 1–100 is a multiple of a given one-digit number. Determine whether a given whole number in the range 1–100 is prime or composite.

Standards for Mathematical PRACTICE

Cool! This is what I'm going to be doing!

1. Make sense of problems and persevere in solving them.
2. Reason abstractly and quantitatively.
3. Construct viable arguments and critique the reasoning of others.
4. Model with mathematics.
5. Use appropriate tools strategically.
6. Attend to precision.
7. Look for and make use of structure.
8. Look for and express regularity in repeated reasoning.

= focused on in this chapter

Am I Ready?

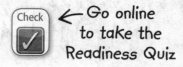 Check ← Go online to take the Readiness Quiz

Multiply. Use models if needed.

1. $2 \times 3 =$ _____

2. $7 \times \$8 =$ _____

3. $9 \times 4 =$ _____

4. $\$7 \times 5 =$ _____

5. Evan's photo album has 8 pages of pictures. How many photos are in Evan's album if the same number of photos is on each page?

_____

Identify the place value of the highlighted digit.

6. 1,630

7. $5,367

8. 20,495

_____   _____   _____

Round each number to its greatest place value.

9. 26 _____

10. $251 _____

11. 4,499 _____

12. There are 1,366 students at Sunrise Elementary School. About how many students attend the school?

_____

Shade the boxes to show the problems you answered correctly.

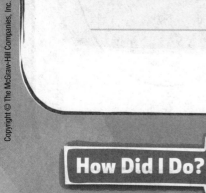 **How Did I Do?**

1	2	3	4	5	6	7	8	9	10	11	12

MY Math Words

Vocab

Review Vocabulary

equation   factor   product

Making Connections

Use the review words to complete each section of the bubble map.
Write an example for or a sentence about each word.

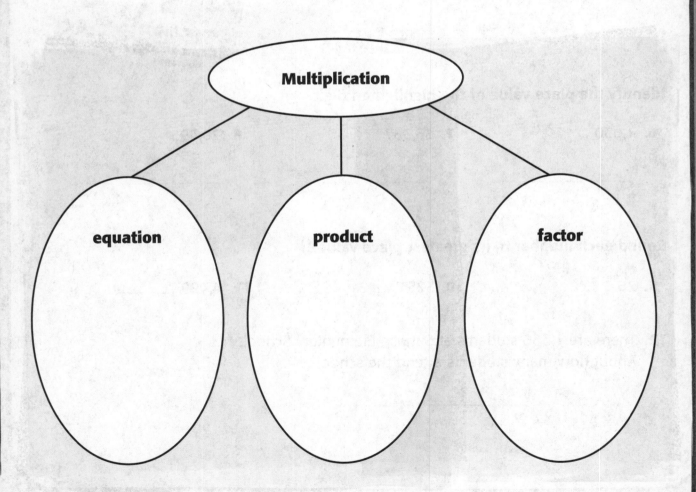

MY Vocabulary Cards

Lesson 4–7

Distributive Property

$$8 \times 11 = (8 \times 10) + (8 \times 1)$$

Lesson 4–4

partial products

$$348 \times 6 = (300 + 40 + 8) \times 6$$
$$300 \times 6 = 1,800$$
$$40 \times 6 = 240$$
$$8 \times 6 = 48$$
$$1,800 + 240 + 48 = 2,088$$

Lesson 4–6

regroup

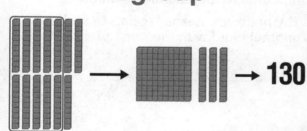
→ 130

Ideas for Use

- During this school year, create a separate stack of cards for key math verbs, such as *regroup*. These verbs will help you in your problem solving.

- Work with a partner to name the part of speech of each word. Consult a dictionary to check your answers.

- Use the blank cards to write your own vocabulary cards.

The products of each place value are found separately, and then added together.

How is using partial products helpful?

_____

_____

_____

Multiply the addends of a number and then add the products.

What are the two operations you use when using the Distributive Property?

_____

To use place value to exchange equal amounts when renaming a number.

The prefix *re-* means "again." Describe another word with the same prefix.

_____

MY Foldable

FOLDABLES® Follow the steps on the back to make your Foldable.

Multi-Digit Multiplication with Zeros

Multiply without Regrouping

Multiply with Regrouping

Multiples

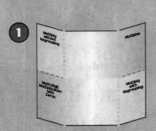
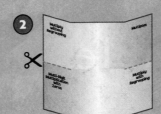
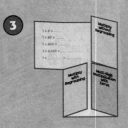

Area Model

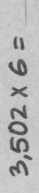

30   2

3

32
x 3
___

7 × 8 = _____

7 × 80 = _____

7 × 800 = _____

7 × 8,000 = _____

7 × _____ = _____

4,238
x   4
_____

3,502 × 6 = _____

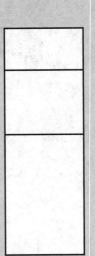

Number and Operations in Base Ten
4.NBT.1, 4.NBT.5, 4.OA.4

CCSS

Multiples of 10, 100, and 1,000

Lesson 1

ESSENTIAL QUESTION
How can I communicate multiplication?

A multiple of a number is the product of that number and any whole number. Any number that is divisible by ten is a multiple of ten. You can use multiples and number patterns to multiply.

 Math in My World Watch Tutor

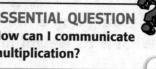 Say "AAAHHH"!

Example 1

The whale shark is the world's largest fish. Its mouth is 5 feet long, and each foot contains 600 teeth. How many teeth does a whale shark have?

Find 5 × 600. Use basic facts and patterns.
600 is a multiple of 10.

5 × 6 = _____   5 × 6 ones = 30 ones = 30

5 × **60** = _____   5 × 6 tens = 30 tens = 300

5 × **600** = _____   5 × 6 hundreds = 30 hundreds = 3,000

So, a whale shark has _____ teeth.

Notice that the product is 5 × 6 with two zeros at the end.

Example 2  Tutor

Find 3 × 7,000.

3 × 7 = _____   3 × 7 ones = 21 ones = 21

3 × **70** = _____   3 × 7 tens = 21 tens = 210

3 × **700** = _____   3 × 7 hundreds = 21 hundreds = 2,100

3 × **7,000** = _____   3 × 7 thousands = 21 thousands = _____

So, 3 × 7,000 is _____.

Notice that the product is 3 × 7 with three zeros at the end.

When you know basic facts and number patterns, you can multiply mentally.

Example 3

The weight of a fire truck is 8 × 4,000 pounds. What is its weight in pounds?

Let *w* represent the weight.

Write an equation.

w = 8 × _____

You need to find 8 × 4,000.

8 × 4 = _____   8 × 4 ones = _____ ones = _____

8 × 40 = _____   8 × 4 tens = _____ tens = _____

8 × 400 = _____   8 × 4 hundreds = _____ hundreds = _____

8 × 4,000 = _____   8 × 4 thousands = _____ thousands = _____

Notice that the product is 8 × 4 with _____ zeros at the end.

Since 8 × 4,000 = _____ , *w* = _____ .

So, the weight of the fire truck is _____ pounds.

Guided Practice

Multiply. Use basic facts and patterns.

1. 6 × 8 = 48   **2.** 7 × 9 = 63

6 × 80 = _____   7 × 90 = _____

6 × 800 = _____   7 × 900 = _____

6 × 8,000 = 48,000   7 × 9,000 = _____

Multiply. Use mental math.

3. 8 × 600 = _____   **4.** 9 × 9,000 = _____

Talk MATH

What is the product of 4 and 5,000? Explain why there are more zeros in the product than in the factors in the problem.

Independent Practice

Multiply. Use basic facts and patterns.

5. $5 \times 3 = 15$
$5 \times 30 = 150$

$5 \times 300 =$ _____

$5 \times 3,000 =$ _____

6. $3 \times 4 = 12$

$3 \times 40 =$ _____

$3 \times 400 =$ _____

$3 \times 4,000 =$ _____

7. $8 \times 5 =$ _____

$8 \times 50 =$ _____

$8 \times 500 =$ _____

$8 \times 5,000 =$ _____

8. $9 \times 1 =$ _____

$9 \times 10 =$ _____

$9 \times 100 =$ _____

$9 \times 1,000 =$ _____

9. $3 \times 7 =$ _____

$3 \times 70 =$ _____

$3 \times 700 =$ _____

$3 \times 7,000 =$ _____

10. $6 \times 5 =$ _____

$6 \times 50 =$ _____

$6 \times 500 =$ _____

$6 \times 5,000 =$ _____

Multiply. Use mental math.

11. $4 \times 30 =$ _____

12. $6 \times 40 =$ _____

13. $7 \times 200 =$ _____

14. $4 \times 500 =$ _____

15. $3 \times 9,000 =$ _____

16. $9 \times 6,000 =$ _____

Algebra Use mental math to find the unknown numbers.

17. If $6 \times 7 = 42$,

then $6 \times$ _____ $= 4,200$.

18. If $5 \times 7 = 35$,

then $5 \times$ _____ $= 3,500$.

19. If $8 \times 3 = 24$,

then $8 \times$ _____ $= 2,400$.

20. If $2 \times 9 = 18$,

then $2 \times$ _____ $= 1,800$.

21. How many times greater is the product 4×300 than the product

$4 \times 30?$ _____

Problem Solving

Weeeeeeee

22. Each ticket for a theme park is $30. What is the total cost for a family with 5 people?

_____

23. The cost for each person to eat for one week is $100. Find the total cost for a family of five to eat for one week.

_____

24. Suppose 5 friends each go on 70 rides. How many rides will they go on altogether?

_____

My Work!

25. Mathematical **PRACTICE 5** **Use Mental Math** Use mental math to find which has a greater product, 5 × 50 or 5 × 500. Explain.

_____

_____

_____

HOT Problems

26. Mathematical **PRACTICE 4** **Model Math** Write two multiplication expressions that have a product of 20,000.

_____

27. **Building on the Essential Question** Does the product of a multiple of 10 always have a zero in the ones place? Explain.

_____

_____

Name _____

MY Homework

Homework Helper

Need help? connectED.mcgraw-hill.com

Find 7 × 5,000.

Use basic facts and patterns to find the product.

7 × 5 = 35   7 × 5 ones = 35 ones, or 35

7 × 50 = 350   7 × 5 tens = 35 tens, or 350

7 × 500 = 3,500   7 × 5 hundreds = 35 hundreds, or 3,500

7 × 5,000 = 35,000   7 × 5 thousands = 35 thousands, or 35,000

So, 7 × 5,000 = 35,000.

Practice

Multiply. Use basic facts and patterns.

1. 4 × 1 = _____

4 × 10 = _____

4 × 100 = _____

4 × 1,000 = _____

2. 6 × 7 = _____

6 × 70 = _____

6 × 700 = _____

6 × 7,000 = _____

3. 3 × 6 = _____

3 × 60 = _____

3 × 600 = _____

3 × 6,000 = _____

4. 8 × 9 = _____

8 × 90 = _____

8 × 900 = _____

8 × 9,000 = _____

Multiply. Use mental math.

5. $2 \times 70 = $ _____

6. $9 \times 500 = $ _____

7. $7 \times 4{,}000 = $ _____

8. $3 \times 2{,}000 = $ _____

Algebra Find each unknown.

9. $30 \times \blacksquare = 120$ _____

10. $6 \times \blacksquare = 3{,}600$ _____

11. $2 \times \blacksquare = 800$ _____

12. $\blacksquare \times 600 = 7{,}200$ _____

Problem Solving

13. Joe bought a house. His payments are $1,000 per month. How much will he pay in 5 months' time?

_____

14. Erin wants to purchase 3 CDs for $10 each. How much money does she need?

_____

15. Kamil makes $100 per week mowing lawns. How much will Kamil make in 6 weeks?

_____

16. **Mathematical PRACTICE** **2** **Use Number Sense** Justin is planning to purchase 1 book for $20 each month for 1 year. How much money will he spend on books in that year?

Test Practice

17. Violet has 11 rolls of pennies. There are 50 pennies in each roll. How many pennies does Violet have altogether?

Ⓐ 5,500 pennies

Ⓑ 550 pennies

Ⓒ 500 pennies

Ⓓ 55 pennies

Round to Estimate Products

Lesson 2

ESSENTIAL QUESTION
How can I communicate multiplication?

To estimate products, you can round factors to their greatest place.

 Math in My World Watch Tutor

Example 1

There are passenger trains in the world that can travel at 267 miles per hour. About how far could a train travel in 3 hours?

Estimate 3 × 267.

Round 267 to its greatest place value.

Then use basic facts and patterns to multiply.

3 × 267

↓

3 × _____ = _____

Helpful Hint

267 rounds to _____.

To which place was 267 rounded? _____

Since 267 was rounded *up*, the estimated product is *greater* than the actual product.

So, the train can travel about _____ miles in _____ hours.

Example 2

Estimate 8 × 2,496.

Round 2,496 to the thousands place value. Then multiply using basic facts and patterns.

8 × 2,496

↓

8 × _____ = _____

Since 2,496 was rounded *down*, the estimated product is *less* than the actual product.

So, 8 × 2,496 is about _____.

Helpful Hint

2,496 rounds

to _____.

Example 3

Lacey's older brother is going to a four-year college. The cost of his tuition is $8,562 each year. About how much will 4 years of college tuition cost?

Estimate 4 × $8,562.
Round to the greatest place value. Then multiply.

4 × _____ = _____

Since 8,562 was rounded _____, the estimated

product is _____ than the actual product.

So, tuition will cost about _____.

College Tuition

Cost per Year

$8,562

Guided Practice

Estimate. Round to the greatest place value. Circle whether the estimate is *greater than* or *less than* the actual product.

Talk MATH

Which product is closer to the estimate of 1,600: 4 × 385 or 4 × 405? Explain.

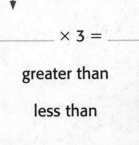

1. 9 × $870

↓

9 × _____ = _____

greater than

less than

2. 3,293 × 3

↓

_____ × 3 = _____

greater than

less than

Independent Practice

Estimate. Round to the greatest place value. Circle whether the estimate is *greater than* or *less than* the actual product.

3. 562 × 6

greater than

less than

_____ × _____ = _____

4. 2 × 896

greater than

less than

_____ × _____ = _____

5. 729 × 8

greater than

less than

_____ × _____ = _____

6. 2 × $438

greater than

less than

_____ × _____ = _____

7. $450 × 7

greater than

less than

_____ × _____ = _____

8. 3 × 5,489

greater than

less than

_____ × _____ = _____

Draw lines to match each product with its most reasonable estimate.

9. 7 × 189

• 4,800

10. 211 × 9

• 1,400

11. 8 × 632

• 2,500

12. 455 × 5

• 1,800

Problem Solving

Toby and Lena like to go to the arcade. They earn points towards prizes. For Exercises 13–15, use the information to the right.

2,000   10,000   50,000   500

13. **Mathematical PRACTICE 1** **Plan Your Solution** Toby went to the arcade 2 times. He earned 5,150 points each time. What is the biggest prize Toby can get?

_____

14. How many toy cars could Toby get with his points?

_____

15. Lena went to the arcade 7 times. She earned 9,050 points each time. What are the two largest prizes she can get?

_____

16. The students in Mrs. Pluma's class each wrote 4 letters to their pen pals. There were about 80 letters written in all. About how many students are in Mrs. Pluma's class?

_____

My Work!

HOT Problems

17. **Mathematical PRACTICE 2** **Use Number Sense** Explain how you can tell if your estimated answer is greater or less than the exact answer to a multiplication problem.

_____

_____

18. **Building on the Essential Question** How is estimation helpful when finding a product mentally? Explain.

_____

MY Homework

Homework Helper

Need help? connectED.mcgraw-hill.com

Estimate 6 × 8,825.

1 Round the greater factor to its greatest place.

Helpful Hint
8,825 rounds up to 9,000

6 × 8,825

↓

9,000

↓

6 × 9,000 = 54,000

2 Multiply. Use basic facts and patterns.
6 × 9,000 = 54,000

The estimate for 6 × 8,825 is 54,000.

Since 8,825 was rounded up, the estimate is greater than the actual product.

Practice

Estimate. Round to the greatest place value. Circle whether the estimate is *greater than* or *less than* the actual product.

1. 756 × 4

800 × 4 = _____

greater than   less than

2. $246 × 8

_____ × 8 = _____

greater than   less than

3. 4,528 × 4

_____ × 4 = _____

greater than   less than

4. 2,331 × 5

_____ × 5 = _____

greater than   less than

Estimate. Round to the greatest place value. Circle whether the estimate is _greater than_ or _less than_ the actual product.

5. 143×2

_____ $\times 2 =$ _____

greater than   less than

6. $2,721 \times 4$

_____ $\times 4 =$ _____

greater than   less than

7. $6 \times \$6,517$

_____

greater than   less than

8. $7 \times \$9,499$

_____

greater than   less than

 # Problem Solving

Estimate each product. Then write whether the estimate is _greater than_ or _less than_ the actual product.

9. There are 62 rows of 9 chairs in the movie theater. About how many chairs are there?

_____

10. Celine is estimating the number of tiles in a mosaic. There are 7 different colors, and 1,725 tiles of each color. What should Celine's estimate be?

_____

11. **Mathematical PRACTICE 5** **Use Mental Math** A resort has 380 rooms with space for 6 people. Estimate the greatest number of people who could be in these rooms at the same time.

_____

Test Practice

12. Kurt swims 575 yards at each swim practice. If he practices Monday, Wednesday, and Friday, about how many yards will he swim that week?

Ⓐ 150 yards   Ⓒ 1,500 yards

Ⓑ 180 yards   Ⓓ 1,800 yards

Hands On
Use Place Value to Multiply

Lesson 3

ESSENTIAL QUESTION
How can I communicate multiplication?

You can use base-ten models to multiply by one-digit numbers.

Build It   Tools

Grace and her friends are at the mall. They see 3 rows of parked cars. There are 23 cars in each row. How many cars are there altogether?

Find 3 × 23 using models.

1 Circle the ones. Count the number of ones.

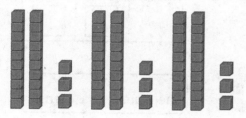

_____ + _____ + _____ = _____

2 Circle the tens. Count the number of tens.

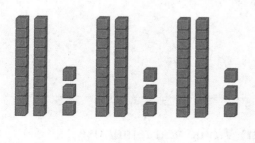

_____ + _____ + _____ = _____

3 × 23 = _____

So, Grace and her friends see _____ cars.

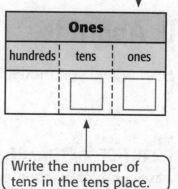

Write the number of ones in the ones place.

Ones		
hundreds	tens	ones

Write the number of tens in the tens place.

Try It

Find 2 × 32 using models.

1 Model 2 × 32.

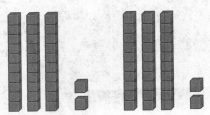

2 Circle the ones. Then count the ones. Complete the place-value chart.

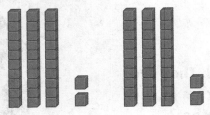

3 Circle the tens. Count the number of tens. Complete the place-value chart.

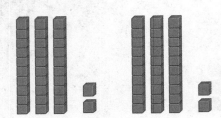

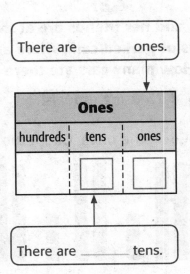

There are _____ ones.

Ones		
hundreds	tens	ones

There are _____ tens.

So, 2 × 32 = _____.

Talk About It

1. How would you model 2 × 22?

_____

2. **Mathematical PRACTICE 3** **Justify Your Conclusion** Would you rather use base-ten blocks, counters, or mental math to model Activity 1? Explain.

_____

_____

Practice It

Multiply. Use models. Draw your models.

3. $3 \times 22 =$ _____

4. $4 \times 12 =$ _____

5. $3 \times 20 =$ _____

6. $1 \times 56 =$ _____

**Algebra Find the unknown number. Use models.
Draw your models.**

7. $4 \times 22 = a$

$a =$ _____

8. $2 \times 24 = c$

$c =$ _____

Apply It

Algebra Write an equation to solve.

9. **Mathematical**
PRACTICE  **Use Symbols** There are 4 shoe stores at the local mall. There are 22 people in each shoe store. How many people are there in the four stores?

My Work!

4 × _____ = _____

So, there are _____ people.

10. Justine earned $32 each month for 3 months. What is the total amount of money that she earned?

_____ × $32 = $_____

So, she earned $_____ .

11. Trent's cousin wants 2 tulips on each table at her wedding. There are 24 tables. How many tulips does she need?

_____ × _____ = _____

So, _____ tulips are needed.

12. **Mathematical**
PRACTICE  **Use Number Sense** The product of two numbers is 88. The sum of the two numbers is 26. What are the numbers?

_____

Write About It

13. How do models help you to multiply by one-digit numbers? Explain.

_____

_____

_____

Number and Operations in Base Ten
4.NBT.5

CCSS

MY Homework

Homework Helper

Need help? connectED.mcgraw-hill.com

Use models to find 4 × 22.

 Count the number of ones.

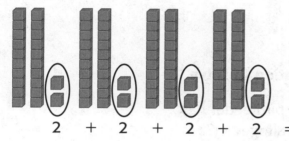

2 + 2 + 2 + 2 = 8

Ones		
hundreds	tens	ones
		8

2 Count the number of tens.

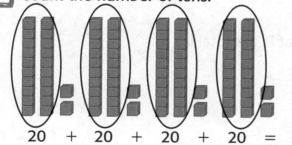

20 + 20 + 20 + 20 = 80

Ones		
hundreds	tens	ones
	8	8

So, 4 × 22 = 88.

Practice

Multiply. Draw the models if needed.

1. 2 × 23 = _____

2. 4 × 21 = _____

3. 2 × 22 = _____

4. 3 × 11 = _____

Multiply. Draw the models if needed.

5. $3 \times 32 =$ _____

6. $2 \times 43 =$ _____

7. $4 \times 12 =$ _____

8. $3 \times 21 =$ _____

My Drawing!

Problem Solving

Mathematical PRACTICE 2 **Use Algebra** Write an equation to solve.

9. Julie hung 3 birdfeeders in her yard. Each birdfeeder has 12 perches. How many perches are there altogether?

_____

10. There are 32 chairs in each classroom. What is the total number of chairs in 2 classrooms?

_____

11. Lance took 34 photos each day of his vacation. He was on vacation for 2 days. Which equation describes how many photos he took in all?

_____

12. There are 42 biscuits in each box. How many biscuits are in 2 boxes?

_____

13. Each prize is worth 3 tickets at the game center. How many tickets are needed for 23 prizes?

Hands On
Use Models to Multiply

Lesson 4
ESSENTIAL QUESTION
How can I communicate multiplication?

You can use **partial products** to multiply a one-digit number by a two-digit number. Find the products for the tens and ones separately. Then add them together.

Area models and arrays can show partial products.

Draw It

The hiking club has 3 groups of hikers. Each group has 12 people. How many hikers are in the hiking club altogether?

Use an array to find 3×12.

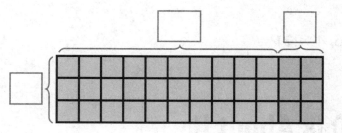

1 Draw a rectangular array.
Separate 12 into

_____ and _____ .

2 Find the partial products.

$3 \times 10 = \boxed{}$

$3 \times 2 = \boxed{}$

3 Add the partial products.

$\boxed{} + \boxed{} = \boxed{}$

$3 \times 12 = $ _____

So, there are _____ hikers in the hiking club.

Online Content at connectED.mcgraw-hill.com

Try It

The hikers hiked for four hours. Each hour, they saw 21 animals. How many animals did they see in all?

Use an area model to find 4 × 21.

 Draw an area model. Separate 21 into

_____ and _____. Label the model.

 Find the partial products.   4 × 20 = _____   4 × 1 = _____

↓   ↓

 Add the partial products.   _____ + _____ = _____

So, 4 × 21 = _____ .

Talk About It

1. Explain how to draw an area model to represent 2 × 15.

_____

_____

_____

2. **PRACTICE** **6** **Explain to a Friend** Explain to a friend how you would multiply 3 × 32.

_____

_____

Name _____

Practice It

Draw an array to multiply.

3. 3 × 13 = _____

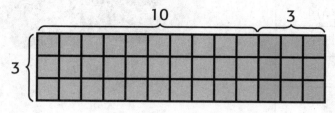

$$3 × 10 = 30$$
$$3 × 3 = 9$$
$$30 + 9 = \underline{\hspace{2cm}}$$

4. 4 × 12 = _____

5. 1 × 26 = _____

Draw an area model to multiply.

6. 3 × 31 = _____

7. 4 × 22 = _____

_____ + _____ = _____   _____ + _____ = _____

Algebra Find each unknown number. Use an array or area model.

8. 43 × 2 = d

d = _____

9. 39 × 1 = g

g = _____

 **Apply It**

Use the table for Exercises 10–13. Draw models to solve.

Toy Sale		
Type of Toy	Regular Price	Sale Price
block sets	$20	$14
puzzles	$12	$10
action figure sets	$13	$12
cars	$11	$10

My Work!

10. What is the total cost of 2 puzzles and 1 car at regular price?

11. How much more are 2 block sets than 2 action figure sets at regular price?

12. **Mathematical PRACTICE** ➊ **Make Sense of Problems** How much is saved by buying 3 action figure sets at the sale price instead of the regular price?

13. **Mathematical PRACTICE** ➋ **Understand Symbols** Compare using the regular prices. Use >, <, or =.

4 cars + 2 action figure sets ◯ 4 puzzles + 1 block set

Write About It

14. How can I use area models to represent multiplication? Explain.

Number and Operations in Base Ten
4.NBT.5
CCSS

MY Homework

Lesson 4

Hands On:
Use Models to
Multiply

Homework Helper

Need help? ⟋ connectED.mcgraw-hill.com

Hailey's parents had a garage sale. Hailey sold 24 games for $2 each. What is the total amount of money Hailey made?

Use an array to find $2 × 24. The array shows 2 rows of 24, which equals 48.

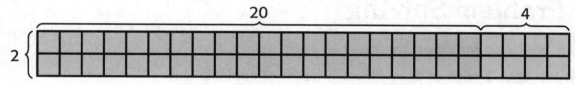

20   4

2 {

1 Find the partial products.

$2 × 20 = 40$
$2 × 4 = 8$

$2 × 24 = 48$

2 Add the partial products

$40 + 8 = 48$

So, Hailey made $48.

Practice

Draw an array or area model to solve.

1. $3 × 22 =$ _____

2. $2 × 41 =$ _____

My Drawing!

Draw an array or area model to solve.

3. $31 \times 3 =$ _____

4. $42 \times 2 =$ _____

5. $24 \times 2 =$ _____

6. $33 \times 2 =$ _____

 Problem Solving

Algebra **Write an equation to solve.**

7. Tyrone uses 2 cups of flour for each batch of cookies. How much flour will he need to make 31 batches of cookies?

_____

8. Corinne's puppy eats 3 meals a day. How many meals does the puppy eat in 32 days?

_____

9. **Mathematical** **PRACTICE** 2 **Use Symbols** There are 3 shelves. Each shelf has 21 books. How many books are there in all?

_____

Vocabulary Check

10. Show how you would use partial products to solve 3×12.

Check My Progress

Vocabulary Check

1. Match each definition to the corresponding vocabulary word.

A(n) _____ of a number is the product of that number and any whole number.	• estimate
A number close to an exact value.	• factor
A number that divides a whole number evenly. Also a number that is multiplied by another number.	• partial products
The products of each place value are found separately, and then added together.	• multiple

Concept Check

Multiply. Use basic facts and patterns.

2. 2 × 60 = _____   **3.** 9 × 600 = _____   **4.** 6 × 4,000 = _____

Estimate. Round to the greatest place value. Circle whether the estimate is *greater than* or *less than* the actual product.

5. 6 × 423

6 × _____ = _____

greater than

less than

6. 1,987 × 5

_____ × 5 = _____

greater than

less than

Draw an array or area model to multiply.

7. 2 × 15 = _____

8. 3 × 19 = _____

Problem Solving

9. Grasshoppers can jump about 20 times their length. About how far could the grasshopper below jump?

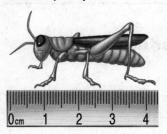

_____

10. There are 21 boxes of markers in the art room. Each box holds 8 markers. How many markers are in the art room?

_____

Test Practice

11. To find the product of 2 × 200, Julia used the basic fact 2 × 2 = 4. How many zeros should she include in the product of 2 × 200?

Ⓐ 1   Ⓒ 3

Ⓑ 2   Ⓓ 4

Multiply by a Two-Digit Number

Place value can help you multiply.

 Math in My World

Example 1

Ann's mom buys two helmets. Each helmet costs $24. How much does she spend on helmets?

You need to find 2 × $24.

Multiply.

$$\begin{array}{r} 24 \\ \times\ 2 \\ \hline 8 \end{array}$$
Multiply the ones.

2 × 4 ones = _____ ones

$$\begin{array}{r} 24 \\ \times\ 2 \\ \hline 48 \end{array}$$
Multiply the tens.

2 × 2 tens = _____ tens

So, Ann's mom spends _____ .

Helpful Hint
Line up the factors by the ones place.

Check for Reasonableness

The area model shows the partial products.

	20	4	
2	2 × 20 = 40	2 × 4 = 8	40 + 8 = ☐

So, the answer is correct.

You can estimate to check for reasonableness.

Example 2

Thirty-one bikes were ordered from Jimmy's Bike Shop. Each bike has two tires. How many tires will Jimmy need for the bikes?

Estimate 31 × 2 ⟶ _____ × _____ = _____

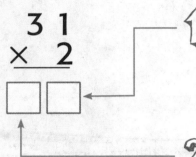

1 Multiply the ones.

2 × 1 one = 2 ones

Write the product in the ones place.

2 Multiply the tens.

2 × 3 tens = 6 tens

Write the product in the tens place.

So, Jimmy will need _____ bike tires.

Check for Reasonableness

The product, _____, is close to the estimate, _____.

Talk MATH

Suppose you found 99 as the product of 33 × 3. How can you check to see if your answer is reasonable?

Guided Practice

Multiply. Check for reasonableness.

1.   4 2
       × 2
    ☐☐

2.   2 1
       × 3
    ☐☐

3. 11 × 4 = _____

4. 32 × 2 = _____

Name ....

CCSS

Independent Practice

Multiply. Check for reasonableness.

5. $\begin{array}{r} 44 \\ \times\ 2 \\ \hline \end{array}$ ☐☐

6. $\begin{array}{r} 21 \\ \times\ 4 \\ \hline \end{array}$ ☐☐

7. $\begin{array}{r} 13 \\ \times\ 2 \\ \hline \end{array}$ ☐☐

8. $41 \times 2 =$ _____

9. $12 \times 3 =$ _____

10. $4 \times 22 =$ _____

Algebra **Find each unknown number.**

11. $41 \times 2 = h$

$h =$ _____

12. $12 \times 3 = j$

$j =$ _____

13. $4 \times 22 = k$

$k =$ _____

Problem Solving

14. **Mathematical PRACTICE 1** **Plan Your Solution** A city has 23 swing sets. Each has 3 swings. How many swings are there in all?

_____

15. The world's largest rodent is called a *capybara*. It can weigh 34 kilograms. How much could 2 capybaras weigh?

_____

16. Recycling one ton of paper saves 17 trees. About how many trees are saved if 4 tons of paper are recycled?

_____

HOT Problems

17. **Mathematical PRACTICE 6** **Explain to a Friend** Adrian has four boxes of action figures. There are 12 in each box. Alec has 21 action figures in each of his three boxes. Who has more action figures? Explain to a friend.

_____

_____

18. **Building on the Essential Question** How can estimation be used to check multiplication problems for reasonableness?

_____

_____

_____

MY Homework

Homework Helper

Need help? connectED.mcgraw-hill.com

Find 4 × 11.

Estimate the product. 4 × 10 = 40

Multiply.

$$\begin{array}{r} 1\ 1 \\ \times\ \ \ 4 \\ \hline 4\ 4 \end{array}$$

1 Multiply the ones.
The product is in the ones place.

2 Multiply the tens.
The product is in the tens place.

So, 4 × 11 = 44.

Check for Reasonableness The product, 44, is close to the estimate, 40.

Practice

Multiply. Draw an area model. Check for reasonableness.

1.   30
        × 2

2.   21
        × 4

3.   86
        × 1

Multiply.

4. $3 \times 31 =$ _____   **5.** $6 \times 11 =$ _____

Problem Solving

6. Caroline makes $3 an hour pet-sitting for the neighbors. Last summer she worked 23 hours. How much money did Caroline earn?

_____

7. Simon has 12 CDs. He burns 3 copies of each. How many CDs did Simon make?

_____

8. The school cafeteria has 4 rows of tables. Each row has 22 seats. How many students can sit in the school cafeteria at the same time?

_____

9. Mathematical **PRACTICE** **2** **Use Number Sense** Steve is playing a memory game with picture cards. He makes 5 rows and puts 11 cards in each row. How many picture cards is Steve using in this game?

_____

Test Practice

10. John wants to buy birthday gifts for 4 friends. He can spend $20 for each gift. How much money can John spend on gifts in all?

Ⓐ $80   Ⓒ $34

Ⓑ $40   Ⓓ $24

My Work!

Hands On
Model Regrouping

Build It

Sometimes you need to regroup to multiply numbers. Regrouping uses place value to exchange equal amounts when renaming a number.

Find 3 × 26.

 Use base-ten blocks to model 3 × 26.

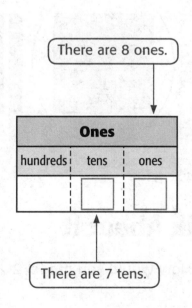

 Find the ones.
There are 18 ones. Regroup as 1 ten and 8 ones.
Write an 8 in the ones place.

Find the tens.
Count the number of tens.
Complete the place-value chart.

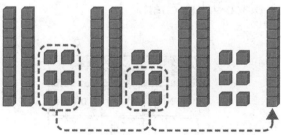

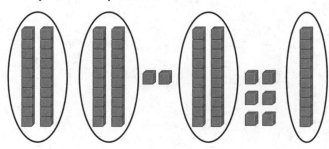

There are 8 ones.

Ones		
hundreds	tens	ones

There are 7 tens.

So, 3 × 26 = _____.

Try It

Find 4 × 31.

1 Use base-ten blocks to model 4 × 31.
Find the ones. Count the ones. There are _____ ones.

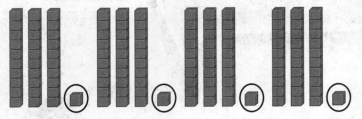

2 Find the tens. There are 12 tens.
Regroup as 1 hundred and _____ tens.

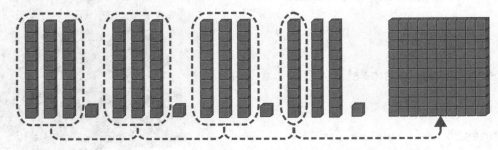

3 Find the hundreds. Count the number of hundreds.
There is _____ hundred. Complete the place-value chart.

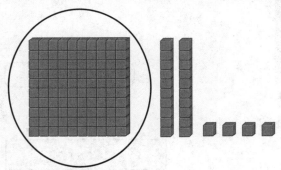

Ones		
hundreds	tens	ones

So, 4 × 31 = _____.

Talk About It

1. How would you model 2 × 38?

Mathematical PRACTICE ➊ Make a Plan How would you model 4 × 52?

Practice It

Multiply. Use models.

3. $2 \times 17 =$ _____

4. $4 \times 32 =$ _____

5. $3 \times 44 =$ _____

6. $4 \times 54 =$ _____

7. $3 \times 28 =$ _____

8. $4 \times 63 =$ _____

9. $2 \times 48 =$ _____

10. $6 \times 24 =$ _____

11. $4 \times 38 =$ _____

12. $5 \times 27 =$ _____

 **Apply It**

13. Willow is shopping for a soccer ball. How many soccer balls *n*, are on the display?

    n = _____ × _____

    There are _____ balls on the display.

14. Suppose there are 6 displays in the store. How many soccer balls *n*, are there in all?

    n = _____ × _____

    There are _____ balls on 6 displays.

15. **Mathematical PRACTICE 2 Use Algebra** Each crate has 35 soccer balls. How many soccer balls *n*, are there in 4 crates?

    n = _____ × _____

    There are _____ balls in 4 crates.

16. **Mathematical PRACTICE 2 Reason** Describe what would happen if you changed the number of balls on the display. How would that change the number of balls on 6 displays?

    _____

    _____

    _____

Write About It

17. How is regrouping for addition similar to regrouping for multiplication?

    _____

    _____

    _____

MY Homework

Homework Helper

Need help? connectED.mcgraw-hill.com

Find 4 × 14.

1 Use base-ten blocks to model 4 × 14.

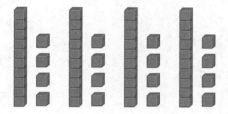

2 Find the number of ones. There are 16 ones. Regroup them as 1 ten and 6 ones.

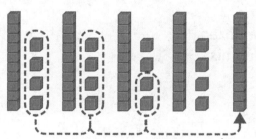

Ones		
hundreds	tens	ones
		6

3 Find the number of tens. There are 5 tens.

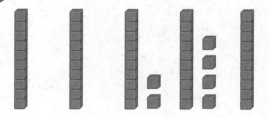

Ones		
hundreds	tens	ones
	5	6

So, 4 × 14 = 56.

Practice

1. Multiply. Draw models if needed. 2 × 46 = _____

Multiply. Draw models if needed.

2. $3 \times 38 =$ _____

3. $4 \times 46 =$ _____

4. $5 \times 23 =$ _____

5. $6 \times 37 =$ _____

Problem Solving

6. Mathematical **PRACTICE** **Use Algebra** Marissa is doing the laundry. She folds 16 items of clothing for each member of her family. There are 6 people in Marissa's family. How many items of clothing does she fold?

_____

7. Ben and his dad sold 9 dozen sunflowers at the farmer's market. There were 12 flowers in one dozen. What is the total number of sunflowers Ben and his dad sold?

_____

8. Riko rides her bike 26 miles every week. What is the total number of miles she will ride in 7 weeks?

_____

My Work!

Vocabulary Check

9. Explain how to regroup 43 ones as tens and ones.

_____

Number and Operations in Base Ten
4.NBT.5
CCSS

The Distributive Property

Lesson 7

ESSENTIAL QUESTION
How can I communicate multiplication?

The **Distributive Property** can be used to multiply greater numbers. It combines addition and multiplication. First, the numbers are decomposed, or broken down into parts. Then, the parts are multiplied separately and then added together.

 Math in My World Watch Tutor

YUM!

Example 1

Chef Cora hard-boils six dozen eggs each day to make egg salad for sandwiches at the food court. How many eggs does she hard-boil each day?

There are 12 eggs in one dozen.

Find 6 × 12.

Decompose 12 into 10 + 2.

Think of 6 × 12 as
(6 × 10) + (6 × 2).

6 × 12 = (_____ × _____) + (_____ × _____)   Decompose 12 into 10 + 2.

= _____ + _____   Find each product mentally.

= _____   Add the products.

So, Chef Cora hard-boils _____ eggs each day.

Key Concept Distributive Property

Words	The Distributive Property says that you can multiply the addends of a number and then add the products.
Symbols	$6 \times 12 = (6 \times 10) + (6 \times 2)$

Example 2

Twenty-seven students are attending a performance at a children's theater. Each admission ticket is $5. What is the total cost for the 27 students?

Let c represent the total. Write an equation.

$c =$ _____ $\times 5$

Find 27×5.

$27 \times 5 = (20 \times 5) + (7 \times 5)$

$=$ _____ $+$ _____

$=$ _____

Since $27 \times 5 =$ _____, $c =$ _____.

So, the total cost for 27 students is _____.

Talk MATH

How can you use the Distributive Property or an area model to find 3×24?

Guided Practice ✓ Check

Use the Distributive Property to multiply. Draw an area model.

1. $12 \times 9 =$ _____ 　　　　 **2.** $22 \times 6 =$ _____

	10	+ 2
9	☐	☐

NARRATOR

Independent Practice

Use the Distributive Property to multiply. Draw an area model.

3.   32
      × 7

4.   15
      × 8

5.   11
      × 8

6.   63
      × 4

7.   55
      × 6

8.   49
      × 9

Algebra Find each unknown number.

9. $37 \times 5 = s$

$s =$ _____

10. $99 \times 9 = t$

$t =$ _____

11. $85 \times 5 = v$

$v =$ _____

12. Write an equation that represents the area model below.

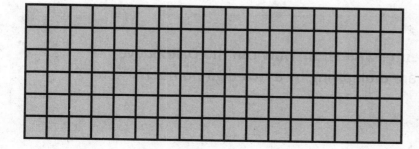

_____ × _____ = _____

Problem Solving

13. Mr. Kline bought 4 sheets of postage stamps. Each sheet had 16 stamps. What is the total number of postage stamps Mr. Kline bought? Write an equation to solve.

$s =$ _____ $\times$ _____

$s =$ _____

Mr. Kline bought _____ stamps.

14. The World of Maps store displays its maps on 3 shelves. There are 26 world maps on each shelf. How many world maps does the store have to sell? Write an equation to solve.

$m =$ _____ $\times$ _____

$m =$ _____

The store has _____ world maps.

My Work!

HOT Problems

15. Mathematical **PRACTICE** ③ **Find the Error** Candace is finding 67×2. She thinks the product is 124. Find and correct her mistake.

_____

_____

_____

16. ❓ **Building on the Essential Question** How can the Distributive Property help when you are multiplying by a two-digit number?

_____

_____

_____

MY Homework

Lesson 7

The Distributive Property

Homework Helper

Need help? ➹ connectED.mcgraw-hill.com

Find 8 × 16.

Think of 8 × 16 as (8 × 10) + (8 × 6).

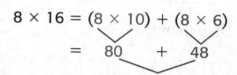

Helpful Hint

Using the Distributive Property, you can multiply the addends of a number. Then add the products.

$8 \times 16 = (8 \times 10) + (8 \times 6)$

$= \quad 80 \quad + \quad 48$

$= \quad\quad 128$

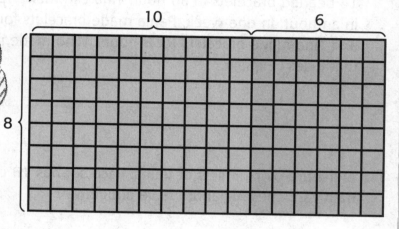

So, 8 × 16 = 128.

Practice

Use the Distributive Property to multiply. Draw an area model.

1. 28
    × 4

2. 19
    × 5

3. 41
    × 6

Multiply. Use the Distributive Property.

4. 75 × 6 = _____   **5.** 4 × 52 = _____   **6.** 8 × 38 = _____

7. 97 × 2 = _____   **8.** 7 × 63 = _____   **9.** 6 × 33 = _____

 # Problem Solving

10. **Mathematical PRACTICE** **Justify Conclusions** Paula can make
14 beaded bracelets in an hour. Tina can make 13 bracelets
in an hour. In one week, Paula made bracelets for 6 hours,
and Tina made bracelets for 8 hours. Who made more bracelets
that week? Explain.

_____

_____

11. Alejandro owns 3 sets of trains. Each set has 18 cars. How many
train cars does Alejandro have altogether?

_____

Vocabulary Check [Vocab]

Write *Distributive Property* or *decompose* on each line.

12. To _____ a number means to break it
into parts.

13. The _____ states that you can multiply
addends of a number, then add the products.

Test Practice

14. Which expression represents the model?

Ⓐ (5 × 10) × (5 × 3)

Ⓑ 5 × 5 × 5 × 3

Ⓒ (5 × 10) + 3

Ⓓ (5 × 10) + (5 × 3)

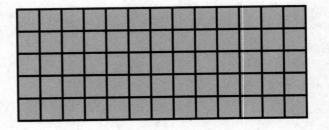

Multiply with Regrouping

Lesson 8

ESSENTIAL QUESTION
How can I communicate multiplication?

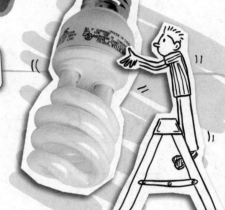

Base-ten blocks can be used to multiply two-digit numbers.

 Math in My World Tools Watch Tutor

Example 1

Zach bought 13 packages of lightbulbs. There are 4 lightbulbs in each package. How many lightbulbs are there in all?

Find 4 × 13.

1 Model 4 groups of 13.

2 Multiply the ones, and regroup.

$$\begin{array}{r} \square \\ 1\ 3 \\ \times\ \ 4 \\ \hline \square\square \end{array}$$

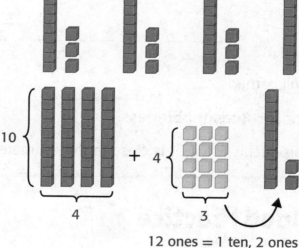

12 ones = 1 ten, 2 ones

3 Multiply the tens. Add the regrouped ten.

So, 4 × 13 = _____.

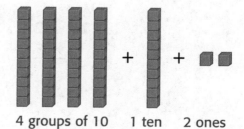

4 groups of 10   1 ten   2 ones

Check Use the Distributive Property.

4 × 13 = (4 × 10) + (4 × 3)

= _____ + _____ = _____

Example 2

Will makes $6 an hour shampooing dogs at a pet shop. Last month he worked 38 hours. How much money did Will earn?

Find 38 × $6.

Estimate   38 × 6 ⟶ _____ × _____ = _____

 Multiply the **ones**.

6 × 8 = 48

$$
\begin{array}{r}
3\,8 \\
\times \quad 6 \\
\hline

\end{array}
$$

 Regroup 48 ones as:

4 tens and

8 ones.

 Multiply the **tens**.

6 × 3 = 18

 Add the regrouped tens.

$$
\begin{array}{r}
18 \\
+ \;\; 4 \\
\hline
22
\end{array}
$$

Talk MATH

Explain how to find 6 × 37.

So, Will earned _____ .

Check for Reasonableness

The product, _____ , is close to the estimate, _____ .

Guided Practice ✓

Multiply. Check for reasonableness.

1. $\begin{array}{r} 23 \\ \times\; 4 \\ \hline \end{array}$ 2. $\begin{array}{r} 42 \\ \times\; 6 \\ \hline \end{array}$

Estimate: Estimate:

_____ × _____ = _____   _____ × _____ = _____

....

Independent Practice

Multiply. Check for reasonableness.

3.   33
    $\times$ 5

4. $24
    $\times$ 4

5.   13
    $\times$ 7

Estimate:

Estimate:

Estimate:

6. $29 \times 4 =$ _____

7. $5 \times 18 =$ _____

8. $7 \times \$36 =$ _____

Estimate:

Estimate:

Estimate:

9. $6 \times 52 =$ _____

10. $75 \times 8 =$ _____

11. $4 \times \$83 =$ _____

Estimate:

Estimate:

Estimate:

Algebra **Find the unknown number in each equation.**

12. $5 \times 31 = x$

13. $63 \times 7 = m$

14. $49 \times 8 = w$

$x =$ _____

$m =$ _____

$w =$ _____

Problem Solving

A local cave has walking tours. Adult tickets cost $18. Child tickets cost $15. Gemstone panning costs $12 per person.

15. The Diaz family has 2 adults and 3 children. How much would it cost for the family to go on a walking tour?

_____

16. Mathematical **PRACTICE** 2 **Reason** Can the Diaz family pan for gemstones for $75? Explain.

_____

_____

17. Find the total cost for the Diaz family to take the walking tour and pan for gemstones.

_____

HOT Problems

18. Mathematical **PRACTICE** 2 **Use Number Sense** Write two multiplication problems that have a product of 120.

_____

19. Mathematical **PRACTICE** 3 **Which One Doesn't Belong?** Circle the multiplication problem that does not belong with the other three. Explain.

12	22	42	33
× 8	× 4	× 2	× 3

20. **?** **Building on the Essential Question** What steps can I use to multiply by a two-digit number with regrouping?

_____

Number and Operations in Base Ten
4.NBT.5, 4.OA.3

CCSS

MY Homework

Lesson 8

Multiply with Regrouping

Homework Helper eHelp

Need help? connectED.mcgraw-hill.com

Find 5 × 37.

Estimate 5 × 40 = 200

Multiply.

$$
\begin{array}{r}
\overset{3}{3}7 \\
\times\ 5 \\
\hline
185
\end{array}
$$

1 Multiply the ones.
5 × 7 = 35
Regroup 35 ones as 3 tens and 5 ones. Write 5 in the ones place. Write 3 above the tens place.

2 Multiply the tens.
5 × 3 = 15
Add the regrouped tens:
15 + 3 = 18
18 tens is 1 hundred and 8 tens.

Check for Reasonableness

The product, 185, is close to the estimate, 200.

You can also use the Distributive Property to check:

(5 × 30) + (5 × 7) = 150 + 35 = 185

Practice

Multiply. Check for reasonableness.

1. $\begin{array}{r} 77 \\ \times\ 3 \\ \hline \end{array}$

2. $\begin{array}{r} \$54 \\ \times\ 6 \\ \hline \end{array}$

3. $\begin{array}{r} 35 \\ \times\ 4 \\ \hline \end{array}$

Estimate: _____

Estimate: _____

Estimate: _____

Multiply. Check for reasonableness.

4. $8 \times \$46 =$ _____

5. $2 \times 93 =$ _____

6. $7 \times 68 =$ _____

Estimate: _____

Estimate: _____

Estimate: _____

7. $4 \times 57 =$ _____

8. $\$7 \times 13 =$ _____

9. $5 \times \$85 =$ _____

Estimate: _____

Estimate: _____

Estimate: _____

 # Problem Solving

10. Mrs. Sands teaches 6 different classes per day at the high school. There are 36 students in each class. How many students does she teach in all?

_____

11. Vin charges $25 rent for each booth at his flea market. If 8 people rent space, how much rent money will Vin collect?

_____

12. **Mathematical PRACTICE 2 Use Number Sense** Becky works 16 days each month. How many days will she work in 6 months?

_____

Test Practice

13. There are 25 gold paper clips and 75 silver paper clips in each box. How many silver paper clips are in 8 boxes?

   Ⓐ 800   Ⓒ 600

   Ⓑ 640   Ⓓ 200

Multiply by a Multi-Digit Number

Lesson 9

ESSENTIAL QUESTION
How can I communicate multiplication?

You can use partial products to multiply by a multi-digit number.

 Math in My World
Watch Tutor

Example 1

Today is Laura's birthday, and she is nine years old. Except for leap years, there are 365 days in one year. How many days old is Laura?

Find 365 × 9.

Estimate 9 × 365 ⟶ 9 × _____ = _____

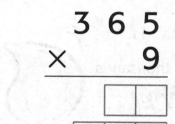

$$
\begin{array}{r}
3\ 6\ 5 \\
\times\ \ \ \ 9 \\
\end{array}
$$

Multiply 9 × 5.

Multiply 9 × 60.

Multiply 9 × 300.

Add the partial products.

	300	+ 60	+ 5
9 {	2,700	540	45

So, Laura is _____ days old.

Check for Reasonableness

The product, _____, is close to the estimate, _____.

Example 2

Find 3 × $1,175.

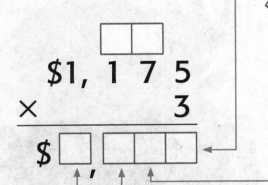

$$\begin{array}{r} \$1,175 \\ \times\ \ \ \ \ 3 \\ \hline \$\Box\,\Box\Box\Box \end{array}$$

1 **Multiply the ones.**

3 × 5 ones = 15 ones

Regroup 15 ones as 1 ten and 5 ones.

2 **Multiply the tens.**

3 × 7 tens = 21 tens

Add the regrouped tens.

21 tens + 1 ten = 22 tens

Regroup 22 tens as 2 hundreds

and 2 tens.

3 **Multiply the hundreds.**

3 × 1 hundred = 3 hundreds

Add the regrouped hundreds.

3 hundreds + 2 hundreds = 5 hundreds

4 **Multiply the thousands.**

3 × 1 thousand = 3 thousands

Talk MATH

Explain why it is a good idea to estimate answers to multiplication problems.

Guided Practice

Multiply. Check for reasonableness.

1. 135
     × 2

2. 532
     × 6

Name ....

Independent Practice

Multiply. Check for reasonableness.

3. 313
    $\times\ 3$

4. 819
    $\times\ 5$

5. $781
    $\times\ 5$

6. 238
    $\times\ 4$

7. $7 \times \$460 =$ _____

8. $7 \times 561 =$ _____

9. $8 \times 6,328 =$ _____

10. $9 \times \$5,679 =$ _____

Algebra Find each unknown number.

11. $8 \times 7,338 = x$

12. $7 \times 8,469 = y$

13. $9 \times \$9,927 = t$

14. $9 \times 8,586 = u$

$x =$ _____

$y =$ _____

$t =$ _____

$u =$ _____

Algebra Find each product if $n = 8$.

15. $n \times 295 =$ _____

16. $737 \times n =$ _____

17. $n \times \$2,735 =$ _____

Compare. Use >, <, or =.

18. 4×198 ◯ 3×248

19. 7×385 ◯ 6×457

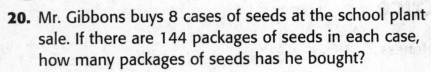
Problem Solving

20. Mr. Gibbons buys 8 cases of seeds at the school plant sale. If there are 144 packages of seeds in each case, how many packages of seeds has he bought?

_____

21. On average, 1,668 gallons of water are used daily by each person in the United States. How much water is used by one person in a week?

_____

22. Each set of furniture costs $2,419. How much would it cost to buy 5 sets of furniture?

_____

My Work!

HOT Problems

23. **Mathematical PRACTICE** ➊ **Keep Trying** Complete the equation.

$$\boxed{}, 287 \times 6 = 25, \boxed{}\, 2\, \boxed{}$$

24. **Mathematical PRACTICE** ➐ **Identify Structure** Write a four-digit number and a one-digit number whose product is greater than 6,000 and less than 6,200.

_____

25. **Building on the Essential Question** How is multiplying by multi-digit numbers similar to multiplying by two-digit numbers?

_____

_____

_____

MY Homework

Lesson 9

Multiply by a Multi-Digit Number

Homework Helper

Need help? connectED.mcgraw-hill.com

Find 3 × 2,763.

Estimate the product: 3 × 3,000 = 9,000

 Multiply the ones.

3 × 3 = 9
Write 9 in the ones place.

 Multiply the tens.

3 × 6 = 18
Regroup 18 tens as 1 hundred and 8 tens.
Write 8 in the tens place.

 Multiply the hundreds.

3 × 7 = 21
Add the regrouped ten.
21 + 1 = 22
Regroup 22 hundreds as 2 thousands
and 2 hundreds.
Write 2 in the hundreds place.

```
    2 1
  2, 7 6 3
×       3
─────────
  8, 2 8 9
```

4 Multiply the thousands.

3 × 2 = 6
Add the regrouped hundreds.
6 + 2 = 8
Write 8 in the thousands place.

Check for Reasonableness

The product, 8,289, is close to the estimate, 9,000.
Another way to check is using partial products.

Multiply 3 × 3.
Multiply 3 × 60.
Multiply 3 × 700.
Multiply 3 × 2,000.
Add the partial products.

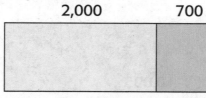

6,000   + 2,100 + 180 + 9 = 8,289

Multiply. Check for reasonableness.

1. 1,313
    × 9

2. $547
    × 6

3. 6,421
    × 3

4. $4,512
    × 5

5. 3,525 × 6 = _____

6. 7 × 7,441 = _____

Algebra Find each product.

7. $n = 8$
   $n \times \$685 =$ _____

8. $n = 3$
   $n \times 5,266 =$ _____

Problem Solving

9. **Mathematical PRACTICE 2 Use Number Sense** One shelf in the greenhouse holds 467 plants. How many plants can 6 shelves hold?

_____

10. Samantha's parents bought her a new bed. They paid $136 each month for 9 months. How much did the bed cost?

_____

11. A concert hall seats 7,689 people. There were 8 concerts in June, and a ticket was sold for every seat. How many tickets were sold in June?

_____

Test Practice

12. Find the product $n \times 2,019$ if $n = 5$.

    Ⓐ 10,000   Ⓒ 10,095

    Ⓑ 10,055   Ⓓ 10,545

Check My Progress

Vocabulary Check

1. Circle the example that correctly shows how to use the **Distributive Property** to find the product of 5 × 15.

   (5 × 10) + (5 × 5)

   (5 × 10) × (5 × 5)

   (5 × 10) × (5 × 15)

   (5 × 10) + (5 × 15)

2. Explain how to use **partial products** to multiply.

   _____

   _____

3. When you use place value to exchange equal amounts when renaming a number, what are you doing?

   _____

Concept Check

Multiply. Check for reasonableness.

4. 23
   × 2

5. 227
   × 8

6. 45
   × 7

7. 612
   × 4

Problem Solving

8. The table shows prices of items at an electronics store.

Electronics Store	
Item	**Price**
Battery pack	$13
Copper wire	$22

How much would it cost to buy 3 battery packs and 3 copper wires?

9. Each campsite needs the number of lanterns shown below. How many lanterns are needed for 48 campsites?

10. There are 1,440 minutes in a day. How many minutes are in 7 days?

Test Practice

11. Mohammed used an area model to show 4 × 35.

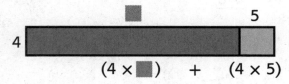

$(4 \times \blacksquare) \quad + \quad (4 \times 5)$

What is the missing number?

Ⓐ 3 Ⓒ 30

Ⓑ 5 Ⓓ 50

My Work!

Problem-Solving Investigation

STRATEGY: Estimate or Exact Answer

Learn the Strategy

There are five girls in Dina's scout troop. They are going to an amusement park. Children's tickets cost $22. What is the total cost of admission? Is an estimate or an exact answer needed?

1 Understand

What facts do you know?

The price of admission is _____ for each girl. There are _____ girls.

What do you need to find?

the total cost of admission and whether an estimate or exact answer is needed

2 Plan

The girls will need to know exactly how much to pay.

So, they need an _____ answer. Find _____ × _____.

3 Solve

$$\begin{array}{r} 22 \\ \times\ 5 \\ \hline \end{array}$$

So, the troop needs _____ to go to the amusement park.

4 Check

Does your answer make sense? Explain.

_____

Practice the Strategy

Each motorized scooter at the scooter store costs $75. About how much would 7 motorized scooters cost? Is an estimate or an exact answer needed?

 Understand

What facts do you know?

_____

_____

What do you need to find?

_____

2 Plan

_____

_____

3 Solve

4 Check

Does my answer make sense? Explain.

_____

Apply the Strategy

Determine whether each problem requires an estimate or an exact answer. Then solve.

My Work!

1. An office needs to buy 6 computers and 6 printers. Each computer costs $384. Each printer costs $88. About $2,400 will be spent on computers. What is the question?

_____

_____

_____

2. Each fourth grade class reads a total of 495 minutes each week. Suppose there are 4 fourth grade classes. How many minutes are read each week?

_____

_____

3. There are 12 stickers on each sheet. There are four sheets in one pack. About how many stickers are in one pack?

_____

_____

4. Mathematical PRACTICE 3 Draw a Conclusion Determine if Tammy, Anessa, and Jaleesa have more than 110 CDs.

Name	CDs Owned
Tammy	21
Anessa	42
Jaleesa	33

_____

Review the Strategies

Use any strategy to solve each problem.
- Use the four-step plan.
- Check for reasonableness.
- Estimate or find an exact answer.

5. Each day, Sparky eats the number of treats shown. How many treats does Sparky eat in one year? (*Hint:* There are 365 days in a year.)

Sparky eats _____ treats in one year.

Mathematical
6. PRACTICE 3 **Find the Error** Each fourth grade class at Tannon Elementary School wants to raise $475 for a local charity. There are 5 fourth grade classes. Rianna says that the overall goal is $2,055. Find and correct her mistake.

_____

_____

_____

7. An electronics store sells remote control dinosaurs for $395 each. About how much do four remote control dinosaurs cost?

_____

_____

8. There are 63 runners in a race. Each runner pays the amount shown to run. How much do the runners pay in all?

$63 \times \$7 =$ _____

My Work!

MY Homework

Homework Helper eHelp

Need help? ➚ connectED.mcgraw-hill.com

Tyrone wants to hand out fliers about a sale to each business in his neighborhood. He plans to leave 3 fliers at each shop, and there are 38 shops. About how many fliers will Tyrone need? Is an estimate or exact answer needed?

1 Understand

What do you know?

Tyrone plans to leave 3 fliers at each shop. There are 38 shops.

What do you need to find?

I need to find about how many fliers Tyrone needs.

2 Plan

The question asks *about* how many fliers are needed. The word *about* means an exact answer is not necessary. The question is asking for an estimate. I will round 38 and multiply.

3 Solve

$$\begin{array}{r} 38 \\ \times\ 3 \end{array} \quad \text{rounds to} \quad \begin{array}{r} 40 \\ \times\ 3 \\ \hline 120 \end{array}$$

So, Tyrone needs about 120 fliers.

4 Check

I will check by comparing the exact answer to the estimate.

$38 \times 3 = (30 \times 3) + (8 \times 3) = 114$

114 is close to the estimate, 120, so the answer is reasonable.

Problem Solving

Determine whether each problem requires an estimate or exact answer. Then solve.

1. Jeff is having a dinner party. He has a large rectangular table that seats 10 people on each long side and 4 people on the two ends. How many people can sit at Jeff's table?

2. Brittany borrowed 3 movies from the library. Each movie is almost 2 hours long. About how many hours of movies does Brittany have to watch?

3. Matt is running the family fun fair at school. He collected about $65 in donations each month for the 7 months he was planning the fair. About how much money does Matt have to spend for the fair?

4. Fatima works at a bakery. She places 5 candied flowers on top of each cupcake she decorates. She will decorate 4 dozen cupcakes today. How many candied flowers will Fatima use today? (*Hint*: 1 dozen = 12)

5. It takes Kayla 12 minutes to walk from her house to her grandparents' house. She walks to and from their house twice a week. About how many minutes each week does Kayla spend walking between the two houses?

6. **Mathematical PRACTICE 2** **Use Number Sense** A haiku is a poem with exactly 3 lines. If each of the 26 students in Mr. Kopp's class writes a haiku, how many lines of poetry does the class write in all?

Name ....

Multiply Across Zeros

Lesson 11

ESSENTIAL QUESTION
How can I communicate multiplication?

You can use the Distributive Property or partial products to multiply across zeros.

 Math in My World Watch Tutor

Example 1

Iván's braces cost $108 each month. How much will Iván's parents pay in 6 months?

Find 6 × $108.

Estimate 6 × $108 ⟶ 6 × _____ = _____

	$100	+	$8
6	6 × $100		6 × $8

6 × 0 = 0, so there is no space in the rectangle for that product.

One Way Distributive Property

6 × $108 = (6 × $100) + (6 × $8)

= (_____) + (_____)

= _____

Another Way Partial Products

$$
\begin{array}{r}
\$108 \\
\times \quad 6 \\
\end{array}
$$

$ ☐☐   6 × $8

$ ☐   6 × $0

+ $ ☐☐   6 × $100

$ ☐☐   Add the partial products.

So, Iván's parents will pay _____ in 6 months.

Check for Reasonableness

The answer, _____, is close to the estimate, _____.

Example 2

If three trees are each 2,025 years old, what is the total age of the trees?

Estimate 3 × 2,025 ⟶ _____ × _____ = _____

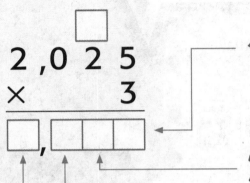

2,025
× 3
_____

1 **Multiply the ones.**

3 × 5 ones = 15 ones
Regroup 15 ones as 1 ten
and 5 ones.

2 **Multiply the tens.**

3 × 2 tens = 6 tens
Add the regrouped tens.
6 tens + 1 ten = 7 tens

3 **Multiply the hundreds.**

3 × 0 = 0 hundreds

4 **Multiply the thousands.**

3 × 2 thousands = 6 thousands

So, the total age of the trees is _____ years.

Check for Reasonableness

The answer, _____, is close to the estimate, _____.

Guided Practice

Multiply. Check for reasonableness.

1. 303
   × 3

2. $507
   × 6

Estimate:

3 × 300 = _____

Estimate:

6 × $500 = _____

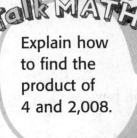

Talk MATH

Explain how to find the product of 4 and 2,008.

Independent Practice

Multiply. Check for reasonableness.

3. 201
$\times$ 2

4. $402
$\times$ 3

5. 709
$\times$ 5

Estimate:

Estimate:

Estimate:

_____ $\times$ _____ = _____   _____ $\times$ _____ = _____   _____ $\times$ _____ = _____

6. $904 \times 9 =$ _____

7. $2 \times \$1,108 =$ _____

8. $4 \times 6,037 =$ _____

Estimate:

Estimate:

Estimate:

9. $8,504 \times 3 =$ _____

10. $6 \times 6,007 =$ _____

11. $5 \times \$9,082 =$ _____

Estimate:

Estimate:

Estimate:

Algebra Find the unknown number.

12. $6 \times 4,005 = s$

$s =$ _____

13. $9,002 \times 9 = q$

$q =$ _____

14. $\$8,009 \times 7 = r$

$r =$ _____

Problem Solving

15. **Mathematical PRACTICE 2** **Stop and Reflect** Large pool equipment sets cost $1,042. Small pool equipment sets cost $907. How much does it cost to buy 3 large pool equipment sets?

_____

How much more does it cost to buy 2 large pool equipment sets than 2 small pool equipment sets?

_____

16. Diller Elementary is collecting money to donate to a charity. $103 is collected each month. How much money is collected in 9 months?

_____

 **HOT Problems**

My Work!

17. **Mathematical PRACTICE 2** **Understand Symbols** Complete the number sentence.

$$\boxed{},\boxed{}\boxed{}\boxed{} \times \boxed{} = \boxed{}\boxed{},\boxed{}\boxed{}\boxed{}$$

18. **Mathematical PRACTICE 3** **Which One Doesn't Belong?** Circle the expression that does not belong. Explain.

| 4,006 × 5 | 3,015 × 2 | 2,010 × 3 | 1,206 × 5 |

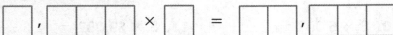

_____

_____

19. **Building on the Essential Question** Why do the products of multi-digit numbers with zeros and one-digit numbers sometimes have zeros in them and sometimes have no zeros in them?

_____

_____

_____

MY Homework

Lesson 11

Multiply Across Zeros

Homework Helper eHelp

Need help? connectED.mcgraw-hill.com

Find 4 × 1,405.

Estimate the product. 4 × 1,000 = 4,000

Multiply.

$$
\begin{array}{r}
\overset{1}{}\ \overset{2}{}\, \\
1{,}405 \\
\times\quad\ 4 \\
\hline
5{,}620
\end{array}
$$

1 **Multiply the ones.**
4 × 5 = 20
Regroup 20 ones as
2 tens and 0 ones.
Write 0 in the ones place.

2 **Multiply the tens.**
4 × 0 = 0
Add the regrouped ones.
0 + 2 = 2
Write 2 in the tens place.

3 **Multiply the hundreds.**
4 × 4 = 16
Regroup 16 hundreds as
1 thousand and 6 hundreds.
Write 6 in the hundreds place.

4 **Multiply the thousands.**
4 × 1 = 4
Add the regrouped hundreds.
4 + 1 = 5
Write 5 in the thousands place.

So, 4 × 1,405 = 5,620.

Check for Reasonableness The product, 5,620, is close to the estimate, 4,000.

Practice

Multiply. Check for reasonableness.

1.
$$
\begin{array}{r}
709 \\
\times\ 6 \\
\hline
\end{array}
$$

2.
$$
\begin{array}{r}
905 \\
\times\ 5 \\
\hline
\end{array}
$$

3.
$$
\begin{array}{r}
5{,}079 \\
\times\ 8 \\
\hline
\end{array}
$$

4.
$$
\begin{array}{r}
2{,}006 \\
\times\ 4 \\
\hline
\end{array}
$$

Multiply. Check for reasonableness.

5. 5,001
    × 9

6. 4,807
    × 7

7. 3,004
    × 3

8. 8,060
    × 3

9. 6,010 × 8 = _____

10. 9,012 × 6 = _____

11. 2 × 1,805 = _____

12. 4 × 1,009 = _____

Problem Solving

13. The art teacher ordered 201 sets of markers for his classroom. Each set has 8 markers. How many markers did he order in all?

14. Brent rode his bicycle 4 miles on Saturday. There are 1,760 yards in a mile. How many yards did Brent ride on Saturday?

15. **Mathematical PRACTICE 2 Use Number Sense** Lorna packs 1,024 socks in each large box at the sock factory. What is the total number of socks in 7 large boxes?

Test Practice

16. There are 405 windows in the office building. Each window has 9 panes. How many window panes does the building have in all?

    (A) 4,059   (C) 3,600

    (B) 3,645   (D) 4,009

Review

Vocabulary Check

Choose the correct word(s) to complete each sentence.
Find the word(s) in the puzzle.

S	T	C	U	D	O	R	P	L	A	I	T	R	A	P	N	F	G	P	D
Y	J	G	S	U	L	X	W	D	M	M	A	R	M	A	A	N	N	R	G
D	P	A	X	Q	T	Y	R	T	Y	U	U	R	P	C	N	T	T	O	P
F	K	E	D	M	T	M	E	N	S	K	R	L	T	U	H	X	P	D	P
F	S	Z	S	T	K	M	I	X	N	I	B	O	T	M	I	S	U	U	N
Q	A	W	D	J	L	O	C	K	O	Z	R	V	V	I	T	B	O	C	O
D	I	S	T	R	I	B	U	T	I	V	E	P	R	O	P	E	R	T	Y
A	H	E	K	C	H	N	K	C	P	A	C	I	P	E	L	L	G	T	C
K	I	H	L	B	K	W	J	A	E	C	R	T	S	X	K	L	E	J	I
W	A	N	A	E	S	T	I	M	A	T	E	B	U	A	Y	P	R	P	E

Distributive Property

estimate

factor

multiple

partial products

product

regroup

1. The answer or result of a multiplication problem is the _____.

2. A(n) _____ is a number close to an exact value.

3. A(n) _____ of a number is the product of that number and any whole number.

4. To _____ is to use place value to exchange equal amounts when renaming a number.

5. The _____ combines multiplication and addition by multiplying each addend by the number and adding the products.

6. To use _____, find the product for the ones, tens, and so forth separately, and then add them together.

7. A number that divides a whole number evenly is called a _____. It is also a number that is multiplied by another number.

Concept Check

Multiply. Use basic facts and patterns.

8. 4 × 90 = _____

9. 6 × 3,000 = _____

Estimate. Round to the greatest place value. Circle if the estimate is _greater than_ or _less than_ the actual product.

10. 1,478 × 4

⬇

_____ × 4 = _____

greater than

less than

11. 5 × 6,225

⬇

5 × _____ = _____

greater than

less than

Multiply. Check for reasonableness.

12.   43
     × 5

13.   24
     × 9

14.   829
      × 8

15.   724
      × 7

16.   569
      × 6

17.   617
      × 3

Problem Solving

18. Marcia read 2 books. Each book was 44 pages long. How many pages did she read?

_____

19. Tania has four decks of 52 cards in each deck. How many cards does Tania have?

_____

20. Each ski pass costs $109. How much do five ski passes cost?

_____

21. A kangaroo can jump as far as 44 feet in a single jump. What distance would three jumps of this size cover?

_____

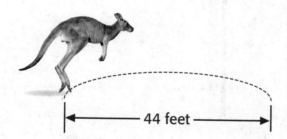

|← 44 feet →|

22. There are 8 party bags. Each bag contains 12 items. Is it reasonable to say that the bags will have 75 items in all? Explain.

_____

My Work!

Test Practice

23. A plane carries 234 passengers. If the plane makes 4 trips a day, how many passengers does the plane transport a day?

Ⓐ 51 passengers   Ⓒ 826 passengers

Ⓑ 800 passengers   Ⓓ 936 passengers

Reflect

Use what you learned about multiplication to complete the graphic organizer.

ESSENTIAL QUESTION

How can I communicate multiplication?

Models

Partial Products

Distributive Property

Example

Reflect on the ESSENTIAL QUESTION Write your answer below.

_____

_____

5 Multiply with Two-Digit Numbers

ESSENTIAL QUESTION

How can I multiply by a two-digit number?

Animals in MY world

Watch a video!

Watch ▶

MY Common Core State Standards

CCSS

Number and Operations in Base Ten

4.NBT.3 Use place value understanding to round multi-digit whole numbers to any place.

4.NBT.4 Fluently add and subtract multi-digit whole numbers using the standard algorithm.

Operations and Algebraic Thinking

4.OA.3 Solve multistep word problems posed with whole numbers and having whole-number answers using the four operations, including problems in which remainders must be interpreted. Represent

4.NBT.5 Multiply a whole number of up to four digits by a one-digit whole number, and multiply two two-digit numbers, using strategies based on place value and the properties of operations. Illustrate and explain the calculation by using equations, rectangular arrays, and/or area models.

This chapter also addresses this standard:

these problems using equations with a letter standing for the unknown quantity. Assess the reasonableness of answers using mental computation and estimation strategies including rounding.

Standards for Mathematical PRACTICE ⬇

Hmm, I don't think this will be too bad!

1. Make sense of problems and persevere in solving them.
2. Reason abstractly and quantitatively.
3. Construct viable arguments and critique the reasoning of others.
4. Model with mathematics.
5. Use appropriate tools strategically.
6. Attend to precision.
7. Look for and make use of structure.
8. Look for and express regularity in repeated reasoning.

= focused on in this chapter

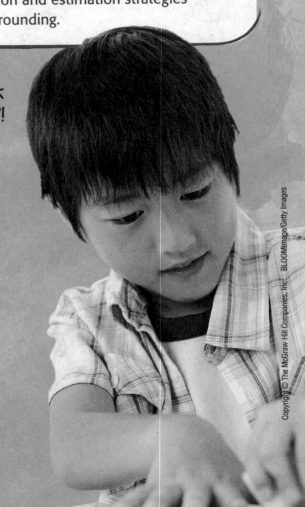

Name _____

Am I Ready?

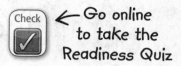
Check ✓ ← Go online to take the Readiness Quiz

Round to the given place.

1. 85,888; nearest ten thousand

_____

2. 681,002; nearest hundred thousand

_____

3. The students raised $6,784 for a new playground. To the nearest thousand, about how much money did the students raise?

_____

Add.

4.
$$\begin{array}{r} 759 \\ + \ 307 \\ \hline \end{array}$$

5.
$$\begin{array}{r} 34,068 \\ + \ 6,055 \\ \hline \end{array}$$

6.
$$\begin{array}{r} 242,607 \\ + \ 480,196 \\ \hline \end{array}$$

Write a multiplication equation that represents each model.

7.

8.

_____ × _____ = _____

_____ × _____ = _____

Multiply.

9. 40 × 9 = _____

10. 36 × 7 = _____

Shade the boxes to show the problems you answered correctly.

How Did I Do? → | 1 | 2 | 3 | 4 | 5 | 6 | 7 | 8 | 9 | 10 |

MY Math Words

Vocab
abc

Review Vocabulary

decompose   equation   factor   product

Making Connections

Use the review vocabulary to complete each section of the bubble
map. Write a sentence about or an example for each word.

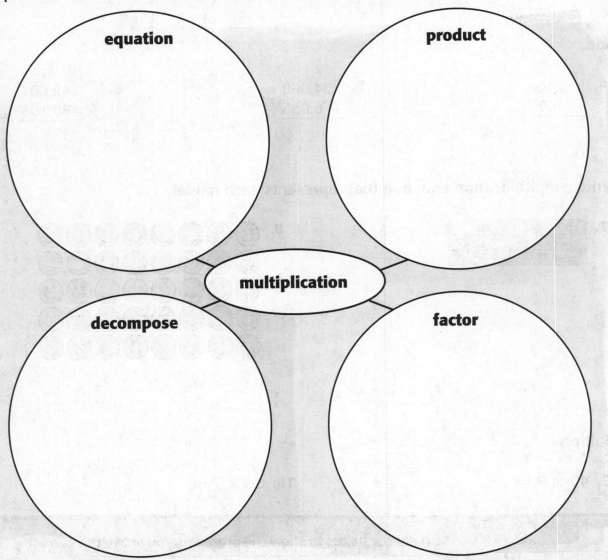

equation

product

multiplication

decompose

factor

Lesson 5–5

operation

$$+ \quad | \quad -$$
$$\times \quad | \quad \div$$

Ideas for Use

- Write a tally mark on each card every time you read the word in this chapter or use it in your writing. Challenge yourself to use at least 10 tally marks for each card.

- Use the blank cards to write your own vocabulary cards.

A mathematical process such as addition, subtraction, multiplication, or division.

Which operations would you use in these equations?

$$200 \bigcirc 4 = 800$$

$$874 \bigcirc 555 = 319$$

MY Foldable

FOLDABLES® Follow the steps on the back to make your Foldable.

ESTIMATE

PRODUCTS

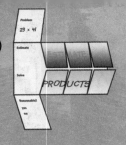
Problem

23×41

Estimate

Solve

Reasonable?

yes

no

Multiply by Tens

Lesson 1

ESSENTIAL QUESTION
How can I multiply by a two-digit number?

 Math in My World [Watch] [Tutor]

Make sure you get my good side!

Example 1

Miss Rita took 20 pictures at the zoo. She printed the pictures so that each of her 25 students could have a copy. How many pictures did Miss Rita print?

Find 25×20.

The number 20 is a multiple of ten.

One Way Use properties.

Think of 20 as _____ × 10.

$25 \times 20 = 25 \times ($ _____ × _____ $)$

$= (25 \times 2) \times 10$

$= 50 \times 10 =$ _____

You have used the Associative Property of Multiplication.

Another Way Use paper and pencil.

1 **Multiply the ones.**

$$\begin{array}{r} 2\ 5 \\ \times\ 2\ 0 \\ \hline \square \end{array}$$

← 0 ones × 25 = _____

2 **Multiply the tens.**

$$\begin{array}{r} 2\ 5 \\ \times\ 2\ 0 \\ \hline \square\ \square\ \square \end{array}$$

← 2 tens × 25 = _____ tens

So, Miss Rita printed _____ pictures.

Example 2

An electronics store has 30 digital music players in stock that cost $99 each. How much do the digital music players cost altogether?

You need to find $99 × 30. The number 30 is a multiple of 10.

 Multiply the ones.

0 ones × 99 = _____

$$\begin{array}{r} \$9\ 9 \\ \times\ 3\ 0 \\ \hline \square \end{array}$$

 Multiply the tens.

3 tens × 99

= _____ tens

$$\begin{array}{r} \$9\ 9 \\ \times\ \ \ 3\ 0 \\ \hline \square\,\square\,\square\,\square \end{array}$$

So, the music players cost a total of $_____.

Helpful Hint
When you multiply a number by a multiple of 10, the digit in the ones place is always zero.

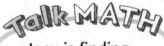

Guided Practice

Multiply.

1.
$$\begin{array}{r} 36 \\ \times\ 10 \\ \hline \end{array}$$

2.
$$\begin{array}{r} 53 \\ \times\ 30 \\ \hline \end{array}$$

3.
$$\begin{array}{r} 42 \\ \times\ 20 \\ \hline \end{array}$$

4.
$$\begin{array}{r} 64 \\ \times\ 40 \\ \hline \end{array}$$

Talk MATH

Joey is finding 67 × 40. Explain why he can think of 67 × 40 as 67 × 4 × 10.

Independent Practice

Multiply.

5. 15
$\times\ 20$

6. 27
$\times\ 30$

7. 46
$\times\ 40$

8. $53 \times 60 =$ _____

9. $80 \times 80 =$ _____

10. $94 \times 90 =$ _____

11. $\$27 \times 10 =$ _____

12. $\$31 \times 30 =$ _____

13. $\$38 \times 50 =$ _____

14. $\$45 \times 50 =$ _____

15. $\$56 \times 70 =$ _____

16. $\$69 \times 80 =$ _____

17. If $7 \times 29 = 203$, then what is 70×29?

_____

18. If $3 \times 52 = 156$, then what is 30×52?

Algebra Use mental math to find the unknown number.

19. $22 \times y = 440$

20. $15 \times y = 450$

21. $25 \times z = 500$

$y =$ _____

$y =$ _____

$z =$ _____

Problem Solving

Hummingbirds feed every 10 minutes. They fly about 25 miles per hour and flap their wings 60 to 80 times each second.

22. What is the least number of times a hummingbird will flap its wings in 15 seconds?

_____

23. What is the greatest number of times a hummingbird will flap its wings in 15 seconds?

_____

24. How many minutes have passed if a hummingbird has eaten 45 times?

_____

25. Mathematical PRACTICE 4 **Model Math** If a hummingbird flies a total of 20 hours, about how far did it fly? Write a number sentence to describe your answer.

HOT Problems

26. Mathematical PRACTICE 3 **Which One Doesn't Belong?** Circle the multiplication problem that does not belong with the other three. Explain.

| 15 × 30 | 28 × 20 | 41 × 21 | 67 × 40 |

_____

27. **Building on the Essential Question** How can place value help me multiply a two-digit number by a multiple of ten?

_____

_____

_____

MY Homework

Homework Helper

Need help? connectED.mcgraw-hill.com

Find 63 × 20.

1 **Multiply the ones.**

0 ones × 63 = 0	→

$$\begin{array}{r} 63 \\ \times\ 20 \\ \hline 0 \end{array}$$

2 **Multiply the tens.**

2 tens × 63 = 126 tens	→

$$\begin{array}{r} 63 \\ \times\ 20 \\ \hline 1{,}260 \end{array}$$

Practice

Multiply.

1.
$$\begin{array}{r} 51 \\ \times\ 30 \\ \hline \end{array}$$

2.
$$\begin{array}{r} 39 \\ \times\ 80 \\ \hline \end{array}$$

3.
$$\begin{array}{r} 25 \\ \times\ 60 \\ \hline \end{array}$$

4.
$$\begin{array}{r} 42 \\ \times\ 50 \\ \hline \end{array}$$

5.
$$\begin{array}{r} 45 \\ \times\ 90 \\ \hline \end{array}$$

6.
$$\begin{array}{r} 88 \\ \times\ 30 \\ \hline \end{array}$$

7. $68 \times 40 = $ _____

8. $11 \times 70 = $ _____

9. $99 \times 10 = $ _____

Problem Solving

10. There are 40 rows of lockers. There are 12 lockers in each row. How many lockers are there?

_____

11. Pablo found out that every classroom has 34 desks. There are 30 classrooms. How many desks are in the school?

_____

12. It costs $10 per person to enter the museum. How much money would it cost for 30 people to enter the museum?

_____

Mathematical
13. **PRACTICE** **6** **Be Precise** Use the Commutative Property to find the unknown number in the equation $35 \times 70 = m \times 35$.

_____

14. **Algebra** Use mental math to find the unknown number in the equation $12 \times b = 480$.

_____

Test Practice

15. To raise money for an animal shelter, 17 students ran in a race. Each student raised $30. How much did the students raise in all?

   Ⓐ $47   Ⓒ $310

   Ⓑ $51   Ⓓ $510

Estimate Products

Lesson 2

ESSENTIAL QUESTION
How can I multiply by a two-digit number?

The word *about* tells you to estimate. When you estimate the product of two 2-digit factors, it is helpful to round them both.

 Math in My World

Example 1

A hamster sleeps 14 hours each day. About how many hours does a hamster sleep in 3 weeks?

There are 21 days in 3 weeks.

So, estimate _____ × _____ .

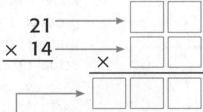 **Round to the nearest ten.**

$$\begin{array}{r} 21 \\ \times\ 14 \end{array}$$

21 rounded to the nearest 10 is _____.

14 rounded to the nearest 10 is _____.

 Multiply.

So, a hamster sleeps about _____ hours in _____ days, or 3 weeks.

Since both factors were rounded down, the estimate is less than the actual product.

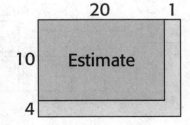

Example 2

Tonya spends 35 minutes playing at the park each day. About how many minutes does she play at the park in 38 days?

You need to estimate 38 × _____ .

 Round each factor to the nearest ten.

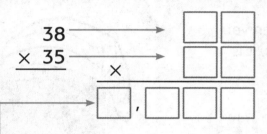

38 rounded to the nearest 10 is _____ .

35 rounded to the nearest 10 is _____ .

 Multiply.

So, Tanya spends about _____ minutes playing at the park.

Since both factors were rounded up, the estimate is _____ than the actual product.

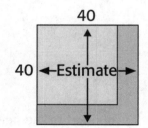

Talk MATH

Explain how you know if an estimated product is greater than or less than the actual product.

Guided Practice

1. Estimate. Circle whether the estimate is *greater than* or *less than* the actual product.

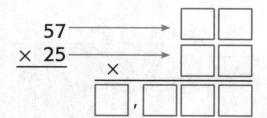

greater than

less than

Independent Practice

Estimate. Circle whether the estimate is *greater than* or *less than* the actual product.

2. 28 →
    × 25 → × _____

greater than

less than

3. 43 →
    × 14 → × _____

greater than

less than

4. $56 →
    × 37 → × _____

greater than

less than

5. 79 →
    × 55 → × _____

greater than

less than

6. $91 →
    × 64 → × _____

greater than

less than

7. 94 →
    × 82 → × _____

greater than

less than

Estimate the product.

8. $23 \times 11 =$ _____

9. $35 \times 37 =$ _____

10. $48 \times 86 =$ _____

11. $53 \times 42 =$ _____

12. $67 \times 56 =$ _____

13. $73 \times 84 =$ _____

Algebra Use mental math to find the unknown number.

14. $20 \times a = 1{,}200$

15. $b \times 30 = 900$

16. $40 \times c = 2{,}400$

$a =$ _____

$b =$ _____

$c =$ _____

Problem Solving

Use the information in the table for Exercises
17–18. Write an equation to solve.

Green Darner Dragonfly Facts	
Average adult length	74 millimeters
Maximum length of nymph	55 millimeters

17. Mathematical PRACTICE 4 Model Math Suppose
18 nymph dragonflies of maximum length are laid
end to end. About how long would they span?

_____ × _____ = _____ mm

18. If 32 average adult dragonflies were laid end to
end, about how long would they span?

_____ × _____ = _____ mm

Algebra Write an equation to solve.

19. The art room has 15 shelves of paint. Each shelf
has 48 cans of paint. About how many cans of
paint are there altogether?

_____ × _____ = _____ cans of paint

20. There are 12 millipedes that each measure
16 centimeters long. About how long would they
measure if they were laid end to end?

_____ × _____ = _____ cm

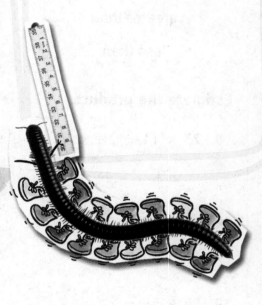

HOT Problems

21. Mathematical PRACTICE 1 Make a Plan Identify two factors
that have an estimated product of 2,000.

_____

22. **Building on the Essential Question** How does an
estimated product relate to the actual product? Explain.

_____

_____

_____

Name

MY Homework

Lesson 2

Estimate Products

Homework Helper

Need help? connectED.mcgraw-hill.com

Estimate 88 × 65. Tell whether the estimate is *greater than* or *less than* the actual product.

1 Round each factor to the nearest ten.

$88 \longrightarrow 90$
$\times 65 \longrightarrow \times 70$

88 rounds to 90.

65 rounds to 70.

2 Multiply.

90
$\times 70$
$\overline{6,300}$

0 ones × 90 = 0

7 tens × 90 = 630 tens

The estimate for 88 × 65 is 6,300.

Since both factors were rounded up, the estimate is greater than the actual product.

Practice

Estimate.

1. 37 × 22 = _____ × _____ = _____

2. 87 × 41 = _____ × _____ = _____

3. 49 × 16 = _____ × _____ = _____

4. 25 × 12 = _____ × _____ = _____

Problem Solving

Estimate. Tell whether the estimate is *greater than* or *less than* the actual product.

5. **Mathematical PRACTICE 4** **Model Math** A concert ticket costs $23. About how much will tickets cost for a group of 22 people?

_____

6. There are 32 times students can work in the computer lab during one week. If 24 students can work in the computer lab at one time, about how many students can work in the computer lab during one week?

_____

Algebra Write an equation to solve.

7. Ramona creates 16 paintings each month. About how many paintings will she create in 3 years?

_____ × _____ = _____ paintings

8. Michael averages 12 points in each basketball game. About how many points will he score in 12 games?

_____ × _____ = _____ points

Test Practice

9. Theater tickets cost $48 per person. About how much money would tickets for 35 people cost?

Ⓐ $2,000   Ⓒ $1,200

Ⓑ $1,500   Ⓓ $200

Check My Progress

Vocabulary Check

1. The **Commutative Property of Multiplication** states that the order in which two numbers are multiplied does not change the product. Provide an example below.

_____

2. The **Associative Property of Multiplication** states that the grouping of factors does not change the product. Provide an example below.

_____

3. An **estimate** is an answer that is close to the exact answer. Provide an example below.

_____

_____

Concept Check

Multiply.

4. $\begin{array}{r} 38 \\ \times\ 30 \\ \hline \end{array}$

5. $\begin{array}{r} 52 \\ \times\ 20 \\ \hline \end{array}$

6. $\begin{array}{r} 47 \\ \times\ 10 \\ \hline \end{array}$

Estimate.

7. $\begin{array}{r} 15 \rightarrow \\ \times\ 28 \rightarrow \\ \hline \end{array}$

8. $\begin{array}{r} 71 \rightarrow \\ \times\ 51 \rightarrow \\ \hline \end{array}$

9. $\begin{array}{r} \$12 \rightarrow \\ \times\ 32 \rightarrow \\ \hline \end{array}$

Problem Solving

10. John jogs 30 miles every week. There are 52 weeks in a year. How many miles does John jog in a year?

_____

11. Ms. Armstrong drives a total of 42 miles each day to and from work. About how many miles does Ms. Armstrong drive in 18 work days?

_____

12. What is the total length of 30 newborn alligators?

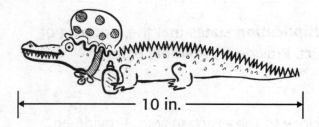

10 in.

_____

13. The average person sends about 25 E-mails a month. About how many E-mails is this each year?

_____

14. Mae is finding the product of 70 × 40. How many zeros will be in the product? Explain.

_____

Test Practice

15. A kangaroo can travel 30 feet per jump. How far could a kangaroo travel if it jumps 14 times?

   Ⓐ 420 feet   Ⓒ 52 feet

   Ⓑ 320 feet   Ⓓ 42 feet

Hands On
Use the Distributive Property to Multiply

Lesson 3
ESSENTIAL QUESTION
How can I multiply by a two-digit number?

You have used the Distributive Property to find a product of a two-digit number and a one-digit number.

$$3 \times 11 = 3 \times (10 + 1)$$
$$= (3 \times 10) + (3 \times 1)$$
$$= \underline{\hspace{2cm}} + \underline{\hspace{2cm}}$$
$$= \underline{\hspace{2cm}}$$

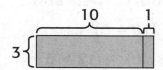

You can also use the Distributive Property to find the product of a two-digit number and a two-digit number.

Draw It

Find 12 × 15.

 Label 12 and 15 as the dimensions of the area model.

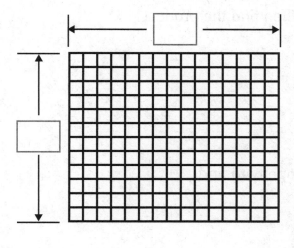

2. Separate the tens and ones of one factor. Label each part.

Write 15 as _____ and _____.

$12 \times 15 = 12 \times (10 + 5)$

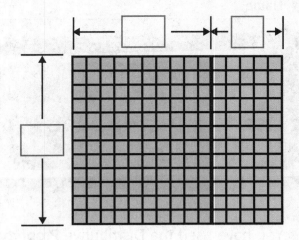

3. Find each product. Then add.

$12 \times 15 = 12 \times (10 + 5)$

$ = (12 \times 10) + (12 \times 5)$

$ = \underline{\hspace{1.5cm}} + \underline{\hspace{1.5cm}}$

$ = \underline{\hspace{1.5cm}}$

	10	5
12	12 × 10 = ☐	12 × 5 = ☐

So, $12 \times 15 = \underline{\hspace{1.5cm}}$.

Talk About It

1. **Mathematical PRACTICE 7 Identify Structure** How would you use the Distributive Property to find 12×18? Then find the product.

_____

_____

2. How would you use the Distributive Property to find 14×17? Then find the product.

_____

_____

Practice It

Draw an area model. Then use the Distributive Property to find each product.

3. Find 36×24.

	20	4
36	720	144

$36 \times 24 = 36 \times (20 + 4)$

$= (36 \times 20) + (36 \times 4)$

$= \underline{\hspace{1.5cm}} + \underline{\hspace{1.5cm}}$

$= \underline{\hspace{1.5cm}}$

4. Find 47×19.

$47 \times 19 = 47 \times (10 + 9)$

$= (47 \times \underline{\hspace{1.5cm}}) + (47 \times \underline{\hspace{1.5cm}})$

$= \underline{\hspace{1.5cm}} + \underline{\hspace{1.5cm}}$

$= \underline{\hspace{1.5cm}}$

5. Find 52×11.

$52 \times 11 = \underline{\hspace{1.5cm}} \times (\underline{\hspace{1.5cm}} + \underline{\hspace{1.5cm}})$

$= (\underline{\hspace{1.5cm}} \times \underline{\hspace{1.5cm}}) +$

$\quad (\underline{\hspace{1.5cm}} \times \underline{\hspace{1.5cm}})$

$= \underline{\hspace{1.5cm}} + \underline{\hspace{1.5cm}}$

$= \underline{\hspace{1.5cm}}$

6. Find 46×22.

$46 \times 22 = \underline{\hspace{1.5cm}} \times (\underline{\hspace{1.5cm}} + \underline{\hspace{1.5cm}})$

$= (\underline{\hspace{1.5cm}} \times \underline{\hspace{1.5cm}}) +$

$\quad (\underline{\hspace{1.5cm}} \times \underline{\hspace{1.5cm}})$

$= \underline{\hspace{1.5cm}} + \underline{\hspace{1.5cm}}$

$= \underline{\hspace{1.5cm}}$

Use the Distributive Property to solve.

7. **Mathematical**
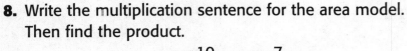
PRACTICE 7 **Identify Structure** There are 15 types of animals in each part of the zoo. The zoo has 12 parts. How many types of animals are there in all?

_____

8. Write the multiplication sentence for the area model. Then find the product.

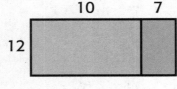

_____ × _____ = _____

9. **Mathematical**
PRACTICE 3 **Find the Error** Tim drew a model to find 11 × 25. Find and correct his mistake.

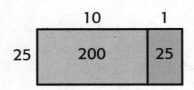

$$200 + 25 = 225$$

_____

_____

Write About It

10. Why is the Distributive Property appropriate for two-digit multiplication? Explain.

_____

_____

My Work!

MY Homework

Homework Helper

Need help? connectED.mcgraw-hill.com

Find 26 × 25.

An area model can be used to represent the factors. The tens and ones of one factor are separated.

Find each product. Then add.

$$26 \times 25 = 26 \times (20 + 5)$$
$$= (26 \times 20) + (26 \times 5)$$
$$= 520 + 130$$
$$= 650$$

So, 26 × 25 = 650.

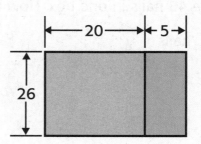

Practice

Draw an area model. Then use the Distributive Property to find each product.

1. 73 × 34 = _____

$$73 \times 34 = 73 \times (30 + 4)$$
$$= (73 \times \rule{1cm}{0.4pt}) + (73 \times \rule{1cm}{0.4pt})$$
$$= \rule{1cm}{0.4pt} + \rule{1cm}{0.4pt}$$
$$= \rule{1cm}{0.4pt}$$

2. 82 × 22 = _____

$$82 \times 22 = 82 \times (20 + 2)$$
$$= (82 \times \rule{1cm}{0.4pt}) + (82 \times \rule{1cm}{0.4pt})$$
$$= \rule{1cm}{0.4pt} + \rule{1cm}{0.4pt}$$
$$= \rule{1cm}{0.4pt}$$

Draw an area model. Then use the Distributive Property to find each product.

3. $18 \times 39 =$ _____ $18 \times 39 =$ _____ $\times ($ _____ $+$ _____ $)$

$= ($ _____ $\times$ _____ $) +$

$($ _____ $\times$ _____ $)$

$=$ _____ $+$ _____

$=$ _____

Problem Solving

4. There are 48 nails in one box. How many nails are in 17 boxes?

_____ nails

$17 \times 48 =$ _____ $\times ($ _____ $+$ _____ $)$

$= ($ _____ $\times$ _____ $) +$

$($ _____ $\times$ _____ $)$

$=$ _____ $+$ _____

$=$ _____

5. Each notebook has 64 pages. How many total pages are there in 33 notebooks?

_____ pages

$33 \times 64 =$ _____ $\times ($ _____ $+$ _____ $)$

$= ($ _____ $\times$ _____ $) +$

$($ _____ $\times$ _____ $)$

$=$ _____ $+$ _____

$=$ _____

6. Each jar contains 55 buttons. There are 16 jars on the shelf. How many buttons are there altogether?

_____ buttons

Multiply by a Two-Digit Number

Lesson 4

ESSENTIAL QUESTION
How can I multiply by a two-digit number?

 Math in My World [Watch] [Tutor]

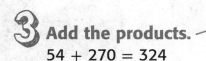 *Are we there yet?*

Example 1

A coyote travels 27 miles each hour. How far can a coyote travel in 12 hours?

Find 27 × 12.

One Way Use partial products.

Draw an area model. Separate each factor into tens and ones. Multiply. Then add the partial products.

200 + 40 + 70 + 14 = _____

	10	2
20	200	40
7	70	14

Another Way Use paper and pencil.

1 Multiply the ones.

$$\begin{array}{r} 27 \\ \times\ 2 \\ \hline 54 \end{array}$$

7 × 2 = 14
Regroup the tens.
2 tens × 2 ones = 4 tens
4 tens + 1 ten = 5 tens

$$\begin{array}{r} 2\ 7 \\ \times 1\ 2 \\ \hline \square\ \square \\ +\ \square\ \square\ \square \\ \hline \square\ \square\ \square \end{array}$$

2 Multiply the tens.
27 × 1 ten = 27 tens, or 270

3 Add the products.
54 + 270 = 324

27 × 12 = _____

So, a coyote can travel _____ miles in 12 hours.

Example 2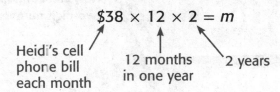

Heidi's monthly bills are shown. How much does she spend on her cell phone service in 2 years?

Write an equation to help you solve the problem.

$$\$38 \times 12 \times 2 = m$$

Heidi's cell phone bill each month

12 months in one year

2 years

Monthly Bills	
Cable	$55
Cell phone	$38
Movie club	$21
Water	$93

You know that $12 \times 2 = 24$. So, you need to find $\$38 \times 24$.

Estimate _____ × _____ = _____

$$\begin{array}{r} \$ 3 \quad 8 \\ \times 2 \quad 4 \end{array}$$

1 Multiply the ones. → $ ☐ ☐ ☐

2 Multiply the tens. → + $ ☐ ☐ ☐

3 Add the products. → $ ☐ ☐ ☐

So, the cost of cell phone service for 2 years is $ _____.

Check

_____ is close to the estimate of _____.

Talk MATH

Explain the steps needed to find the product of 56 and 23.

Guided Practice

Multiply.

1.
$$\begin{array}{r} 3 \quad 5 \\ \times 2 \quad 4 \end{array}$$

☐ ☐ ☐   Multiply the ones.

+ ☐ ☐ ☐   Multiply the tens.

☐ ☐ ☐   Add.

	← 30 →	← 5 →
20	600	100
4	120	20

The area model shows that

$600 + 120 + 100 + 20 =$ _____.

Independent Practice

Multiply. Use the area model to check.

2. 19
    × 15

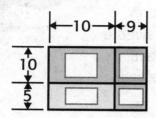

3. 42
    × 38

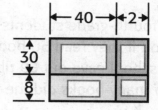

4. $54
    × 51

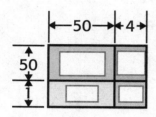

5. $74
    × 63

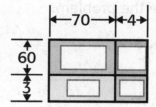

Multiply.

6.   47
    × 24

7.   64
    × 46

8. 83 × 67 = _____

9. 91 × 78 = _____

Problem Solving

10. A greyhound dog can jump a distance of 27 feet. How many feet will a greyhound travel if it jumps 12 times?

_____

11. The fourth-grade students at Tremont School received a ribbon if they read 50 books during the school year. The school gave out 69 ribbons at the end of the year. How many books did the students read in all?

_____

Mathematical
12. PRACTICE 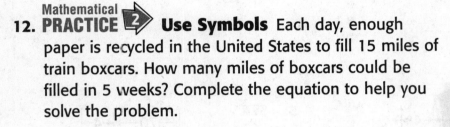 **Use Symbols** Each day, enough paper is recycled in the United States to fill 15 miles of train boxcars. How many miles of boxcars could be filled in 5 weeks? Complete the equation to help you solve the problem.

$$15 \times 5 \times \underline{\qquad} = b$$

_____

My Work!

HOT Problems

Mathematical
13. PRACTICE 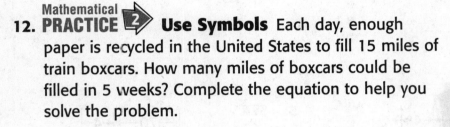 **Which One Doesn't Belong?** Circle the multiplication problem that does not belong with the other three. Explain.

22	$45	37	$66
×15	× 28	× 18	× 25

_____

14. **Building on the Essential Question** Why can't the product of two 2-digit numbers ever be two digits? Explain.

_____

_____

MY Homework

Homework Helper

 eHelp

Need help? connectED.mcgraw-hill.com

Find 29 × 56.

1 Multiply the ones.

2 Multiply the tens.

3 Add the products.

$$
\begin{array}{r}
56 \\
\times\ 29 \\
\hline
504 \\
+\ 1{,}120 \\
\hline
1{,}624
\end{array}
$$

504 ← 9 × 56
1,120 ← 20 × 56

	← 50 →	← 6 →
20	1,000	120
9	450	54

So, 29 × 56 = 1,624.

Practice

Multiply.

1.
$$
\begin{array}{r}
26 \\
\times\ 35 \\
\hline
\end{array}
$$

2.
$$
\begin{array}{r}
\$46 \\
\times\ 35 \\
\hline
\end{array}
$$

3.
$$
\begin{array}{r}
79 \\
\times\ 73 \\
\hline
\end{array}
$$

4.
$$
\begin{array}{r}
73 \\
\times\ 51 \\
\hline
\end{array}
$$

5.
$$
\begin{array}{r}
59 \\
\times\ 47 \\
\hline
\end{array}
$$

6.
$$
\begin{array}{r}
94 \\
\times\ 61 \\
\hline
\end{array}
$$

Multiply.

7.
$$\begin{array}{r} 44 \\ \times\ 87 \\ \hline \end{array}$$

8.
$$\begin{array}{r} 77 \\ \times\ 22 \\ \hline \end{array}$$

Problem Solving

9. Mrs. Taylor gave each of her students 75 pieces of paper at the beginning of the school year. If there are 32 students in her class, how many pieces of paper did she give out altogether?

10. Mr. Matthews gives each of his 32 students 15 minutes to present a book report to the class. How many minutes will it take for all of his students to present their reports?

11. **Mathematical** **PRACTICE** **2** **Use Symbols** George collects 25 baseball cards each month. How many cards will he have at the end of one year? Complete the equation to help you solve the problem.

$25 \times$ _____ $= n$

12. The office building is 48 floors high. Each floor has 36 windows. How many windows does the building have in all?

Test Practice

13. There are 26 rows of bleachers in the school gym. At the pep rally, there were 17 students sitting in each row. How many students were there in all?

Ⓐ 43 students   Ⓒ 208 students

Ⓑ 182 students   Ⓓ 442 students

Number and Operations in Base Ten
4.NBT.4, 4.NBT.5, 4.OA.3

CCSS

Solve Multi-Step Word Problems

Lesson 5

ESSENTIAL QUESTION
How can I multiply by a two-digit number?

Sometimes, problems require more than one operation to solve. An **operation** is a mathematical process such as addition, subtraction, multiplication, or division.

Let's go!

Math in My World

Example 1

Francis earns $8 each week walking dogs. She spends $3 each week and saves the rest. There are 52 weeks in a year. How much will she have saved at the end of the year?

You need to find ($8 − $3) × 52. The operations that are needed for this problem are subtraction and multiplication.

Estimate ($8 − $3) × 52   Round 52 to 50.

$5   × 50 = $250

1 Subtract. ($8 − $3) × 52

$5   × 52

2 Multiply.

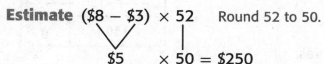

```
      5  2
   ×     5
   ☐ ☐ ☐
```

So, Francis will have saved $ _____ .

Check

$ _____ is close to the estimate $250. So, the answer is correct.

You can use a variable to represent unknown quantities.

Example 2

Coach Murphy bought three boxes of trophies. Each box has 45 soccer trophies. He also bought 15 tennis trophies and some golf trophies. There are a total of 170 trophies. Write an equation to describe the number of trophies that Coach Murphy bought. How many golf trophies did Coach Murphy buy?

Write an equation.

soccer trophies   tennis trophies   golf trophies

$(3 \times 45) + 15 + ? = 170$ trophies

Replace the unknown quantity with a variable.

$(3 \times 45) + 15 + g = 170$

Solve for the unknown quantity.

 Multiply.

$(3 \times 45) + 15 + g = 170$

$\boxed{} + 15 + g = 170$

Add.

$135 + 15 + g = 170$

$\boxed{} + g = 170$

Subtract.

$150 + g = 170$

Subtract 150 from 170 to find the value of g.

$170 - 150 = \underline{}$

170	
150	**g**

So, Coach Murphy bought _____ golf trophies.

Guided Practice

1. Karrie has 20 bags of prizes. Each bag has 4 prizes. She also has a red bag with 13 prizes and a blue bag with the rest of the prizes. She has a total of 100 prizes. How many prizes are in the blue bag? Write an equation. Use a variable for the unknown number.

_____

_____

Talk MATH

Why are variables used?

Name ....

Independent Practice

Algebra Write an equation for each problem. Solve.

2. Each small dog at doggy day care weighs 35 pounds. Each large dog weighs 60 pounds. There are 4 small dogs and 6 large dogs. How much do the dogs weigh altogether?

My Work!

3. Suzie has track practice for 1 hour on Tuesday and 2 hours on Thursday. How many hours does Suzie go to track practice in 15 weeks?

Algebra Write an equation for each problem. Use a variable for the unknown number. Solve.

4. Warren, Lisa, and Tina went to the carnival. The table shows the number of points Warren won at each carnival game.

Game	Points
Banana Bowling	24
Toss & Trade	16
Rabbit Race	10

Lisa won the same number of points as Warren. Warren, Lisa, and Tina won a total of 225 points. How many points did Tina win?

5. Mark bought four hats that each cost $8. He also bought a shirt that cost $14 and a pair of jeans. He spent a total of $68. How much did the jeans cost?

Problem Solving

Use a number cube to complete each number puzzle.

6. **Mathematical PRACTICE**  **Keep Trying** Roll a number cube four times.

The numbers rolled are: _____,

_____, _____, and _____.

Write the numbers in the boxes below. Use each number once. Try to create the greatest number possible.

(☐ × ☐) + ☐ − ☐ = _____

7. Roll a number cube four times.

The numbers rolled are: _____, _____, _____,

and _____.

Write the numbers in the boxes below. Use each number once. Try to create the greatest number possible.

☐ + (☐ × ☐) − ☐ = _____

HOT Problems

8. **Mathematical PRACTICE** **Make Sense of Problems** A bus has 15 rows of seats. Each row has 4 seats. At the first stop, 25 people get on the bus. At the second stop, 3 people get off of the bus, and 12 people get on the bus. How many empty seats are there after the second stop?

_____

_____

9. ❓ **Building on the Essential Question** How can I use equations to model real-world problems?

_____

_____

_____

MY Homework

Homework Helper

Need help? connectED.mcgraw-hill.com

A store had 3 baskets. Each basket had 62 necklaces. In the morning, 25 necklaces were sold. In the afternoon, some necklaces were returned. At the end of the day, there were 166 necklaces. How many necklaces were returned?

Write an equation to represent the problem.

baskets of necklaces	−	necklaces sold	+	necklaces returned	= total
↓		↓		↓	↓
(3×62)	−	25	+	n	= 166

Use a variable.

Solve for the unknown quantity.

1 **Multiply.**

$(3 \times 62) - 25 + n = 166$

$\boxed{} - 25 + n = 166$

2 **Subtract.**

$186 - 25 + n = 166$

$\boxed{} + n = 166$

3 **Use mental math.** $161 + n = 166$

$161 + 5 = 166$

So, 5 necklaces were returned.

Practice

1. Gina works at a diner. She earns $6 each hour plus tips. In one week, she worked 37 hours a week and earned $43 in tips. How much did she make altogether? Write an equation. Use a variable for the unknown. Solve.

_____

Problem Solving

Write an equation for each problem. Use a variable for the unknown number. Solve.

2. **Mathematical PRACTICE** **2** **Use Number Sense** It costs $45 to rent a car each day. There is also a fee of $12. How much does it cost to rent a car for 5 days, including the fee?

_____

3. A climbing gym charges $10 to climb each day. A pair of climbing shoes costs $84. It costs $169 to buy 6 days of climbing, one pair of climbing shoes, and one harness. How much does a harness cost?

_____

4. A travel agency charges $64 for each bus ticket and $82 for each train ticket. How much does it cost to buy 3 bus tickets and 4 train tickets?

_____

Vocabulary Check

5. Label each part of the equation. Write *operation* or *variable*.

$$n - 100 + r = 54$$

[_____]   [_____]   [_____]

Test Practice

6. There are three shelves. Each shelf has 28 books. There is also a stack of some more books. There are a total of 85 books. Which equation represents this situation?

Ⓐ $(3 \times 28) + b = 85$ Ⓒ $(3 \times 28) + 85 = b$

Ⓑ $(3 + 28) \times b = 85$ Ⓓ $(3 + 28) \times 85 = b$

Problem-Solving Investigation

STRATEGY: Make a Table

Lesson 6

ESSENTIAL QUESTION
How can I multiply by a two-digit number?

Learn the Strategy

Each roller coaster car holds 18 people. Every minute, a new car is filled. Make a table to find how many people can ride it in 60 minutes.

AAAAHHHH!!!

1 Understand

What facts do you know?

There are _____ people per car.

What do you need to find?

the number of _____ who can ride in _____ minutes

2 Plan

I can make a table to find the number of people who can

ride in _____ minutes.

3 Solve

Start by finding the product of 18 and 10. $18 \times 10 = 180$

Minutes	10	20	30	40	50	60
Passengers	180					

So _____ people can ride the roller coaster in 60 minutes.

4 Check

Does your answer make sense? Explain.

Yes. _____ × _____ = _____

Online content at connectED.mcgraw-hill.com

Practice the Strategy

There are 20 sea lions in a circus. Each sea lion can juggle 5 balls at a time. How many balls will the sea lions need for their act if they all perform at the same time?

Understand

What facts do you know?

_____

_____

What do you need to find?

_____

_____

2 Plan

_____

_____

3 Solve

_____

_____

4 Check

Does your answer make sense? Explain.

_____

Apply the Strategy

Solve each problem by making a table.

1. A page from Dana's album is shown. Dana puts the same number of stickers on each page. She has 30 pages of stickers. How many stickers does she have in all?

Dana has _____ stickers in all.

2. West Glenn School has 23 students in each class. There are 6 fourth grade classes. About how many fourth grade students are there in all?

There are about _____ students. I solved this by

_____

_____

_____

3. **Mathematical** **PRACTICE** **3** **Draw a Conclusion** Evita completed 30 problems for her math homework each night. She has math homework five nights a week. Write a real-world problem using this information. Then solve.

_____

_____

_____

4. Ling exercises for 30 minutes 2 times a day. If she keeps up this schedule for 30 days, how many minutes will she exercise in all?

_____

Review the Strategies

Use any strategy to solve each problem.
- Make a table.
- Find an estimate or exact answer.
- Find reasonable answers.
- Draw a diagram.

5. Corey and his 2 friends earn $12 each for doing yard work. How much money will they get paid altogether if they work on 5 yards? Make a table.

3 friends × $12 a yard = $36

1 yard	2 yards			
$36				

_____

6. A lemur sleeps 16 hours each day. A sloth sleeps 4 more hours each day than a lemur. How many total hours of sleep do a lemur and a sloth get during two days?

_____

7. A lizard eats 6 crickets each day. How many crickets does it eat in 13 weeks?

_____

8. Mathematical PRACTICE 5 Use Math Tools Write a real-world problem that would involve making a table to find the answer.

_____

_____

_____

9. Pete spends 30 minutes a night reading. How many hours does he spend reading in a 30-day month?

_____

My Work!

MY Homework

Homework Helper

Need help? ⟋ connectED.mcgraw-hill.com

The cafeteria serves breakfast to 48 students each morning. In a week of 5 school days, how many times is breakfast served?

1 Understand
I know that 48 students are served breakfast each morning for 5 days.

2 Plan
I can make a table to find 48 × 5.

3 Solve

Day	1	2	3	4	5
Breakfast served	48	96	144	192	240

So, breakfast is served 240 times in one week.

4 Check
Multiply 5 × 48.
5 × 48 = 240

Problem Solving

1. Mrs. Shelley's class is reading *The Lion, the Witch, and the Wardrobe.* If they read 16 pages every week, how many pages can they read in 5 weeks? Solve the problem by making a table.

Week					
Pages					

Solve each problem by making a table.

2. **Mathematical PRACTICE** **8** **Look for a Pattern** Fiona finds 1 shell on her first day at the beach. Every day that week, she finds twice as many shells as she did the day before. How many shells does Fiona find on the seventh day?

3. The parking garage holds 300 cars on each level. There are 4 levels in the garage. How many cars can the parking garage hold in all?

4. Jonah sets the table for breakfast and dinner every Monday, Wednesday, and Friday. How many times does he set the table in six weeks?

5. Tony and his brother have 20 thank-you cards to write. If they each write 2 per day, how many days will it take them to finish?

6. At a dance recital, each dancer performs for 13 minutes. If there are 6 dancers to perform, how long is the recital?

Dancers					
Minutes					

Vocabulary Check

Using the word bank below, write the correct word in each blank.

Associative Property   **Commutative Property**   **Distributive Property**

operation   **partial products**

1. This property states that the order in which two numbers are multiplied does not change the product.

$$23 \times 11 = 11 \times 23$$

_____

2. In the equation $32 \times 10 = 320$, the multiplication symbol represents this process.

_____

3.
```
    18
 ×  27
 ─────
   126
 + 360
 ─────
   486
```

4. This property states that the grouping of the factors does not change the product.

_____

5. This property states that multiplying a sum by a number is the same as multiplying each addend by the number and then adding the products.

$$2 \times 12 = 2 \times (10 + 2)$$
$$= (2 \times 10) + (2 \times 2)$$
$$= 20 + 4$$
$$= 24$$

Concept Check ✓

Multiply.

6.
$$\begin{array}{r} 90 \\ \times\ 90 \\ \hline \end{array}$$

7.
$$\begin{array}{r} 34 \\ \times\ 80 \\ \hline \end{array}$$

8.
$$\begin{array}{r} \$28 \\ \times\ 40 \\ \hline \end{array}$$

9.
$$\begin{array}{r} \$45 \\ \times\ 30 \\ \hline \end{array}$$

Estimate. Circle whether the estimate is *greater than* or *less than* the actual product.

10.
$$\begin{array}{r} \$24 \longrightarrow \\ \times\ 31 \longrightarrow \\ \hline \end{array} \times \underline{\hspace{1.5cm}}$$

greater than

less than

11.
$$\begin{array}{r} 48 \longrightarrow \\ \times\ 89 \longrightarrow \\ \hline \end{array} \times \underline{\hspace{1.5cm}}$$

greater than

less than

12.
$$\begin{array}{r} 37 \longrightarrow \\ \times\ 66 \longrightarrow \\ \hline \end{array} \times \underline{\hspace{1.5cm}}$$

greater than

less than

13.
$$\begin{array}{r} \$52 \longrightarrow \\ \times\ 84 \longrightarrow \\ \hline \end{array} \times \underline{\hspace{1.5cm}}$$

greater than

less than

Multiply.

14.
$$\begin{array}{r} 63 \\ \times\ 46 \\ \hline \end{array}$$

15.
$$\begin{array}{r} 26 \\ \times\ 34 \\ \hline \end{array}$$

16.
$$\begin{array}{r} \$72 \\ \times\ 49 \\ \hline \end{array}$$

17.
$$\begin{array}{r} \$55 \\ \times\ 41 \\ \hline \end{array}$$

Don't get cold feet!

Problem Solving

18. A football coach orders 30 jerseys for his football team. The jerseys cost $29 each. What is the total cost of the jerseys?

_____

19. Julio scores 18 points in each basketball game. If there are 14 games in a season and Julio continues to score 18 points each game, how many points will Julio score?

_____

20. There are 30 students in each class. There are 27 classrooms. How many students are there altogether?

_____

21. Tamara makes $12 an hour. She worked 28 hours this week. About how much money will she make?

_____

22. Austin's allowance is $15 per week. He spends $4 each week on baseball cards. How much money will Austin have at the end of 12 weeks?

_____

Test Practice

23. A jumbo can of vegetables has 36 servings. How many servings of vegetables are in 18 cans?

Ⓐ 648 servings   Ⓒ 608 servings

Ⓑ 324 servings   Ⓓ 54 servings

My Work!

Use what you learned about multiplying with two-digit numbers to complete the graphic organizer.

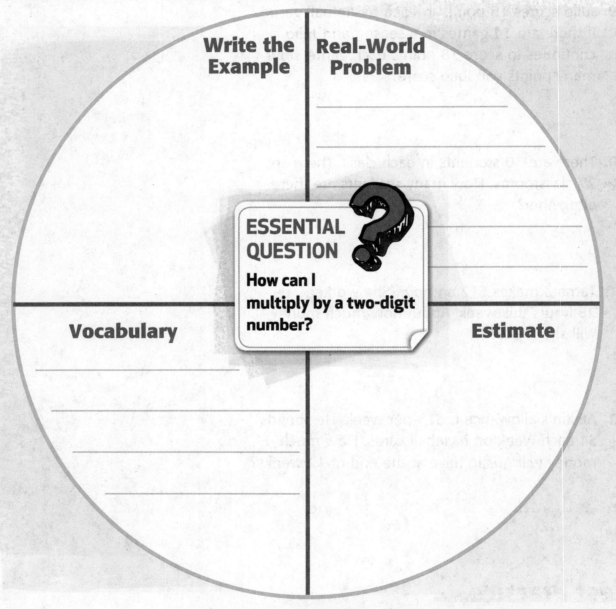

Write the Example

Real-World Problem

ESSENTIAL QUESTION

How can I multiply by a two-digit number?

Vocabulary

Estimate

Reflect on the **ESSENTIAL QUESTION** **Write your answer below.**

6 Divide by a One-Digit Number

Let's Travel!

Watch

Watch a video!

MY Common Core State Standards

Number and Operations in Base Ten

4.NBT.1 Recognize that in a multi-digit whole number, a digit in one place represents ten times what it represents in the place to its right.

4.NBT.3 Use place value understanding to round multi-digit whole numbers to any place.

Operations and Algebraic Thinking *This chapter also addresses these standards:*

4.OA.3 Solve multistep word problems posed with whole numbers and having whole-number answers using the four operations, including problems in which remainders must be interpreted. Represent these problems using equations with a letter standing for the unknown quantity. Assess the reasonableness of answers using mental computation and estimation strategies including rounding.

4.NBT.6 Find whole-number quotients and remainders with up to four-digit dividends and one-digit divisors, using strategies based on place value, the properties of operations, and/or the relationship between multiplication and division. Illustrate and explain the calculation by using equations, rectangular arrays, and/or area models.

4.OA.4 Find all factor pairs for a whole number in the range of 1-100. Recognize that a whole number is a multiple of each of its factors. Determine whether a given whole number in the range 1-100 is a multiple of a given one-digit number. Determine whether a given whole number in the range 1-100 is prime or composite.

Standards for Mathematical PRACTICE

1. Make sense of problems and persevere in solving them.
2. Reason abstractly and quantitatively.
3. Construct viable arguments and critique the reasoning of others.
4. Model with mathematics.
5. Use appropriate tools strategically.
6. Attend to precision.
7. Look for and make use of structure.
8. Look for and express regularity in repeated reasoning.

= focused on in this chapter

Ok, this'll be good to know!

Name _____

Am I Ready?

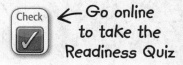

Check ✓ ← Go online to take the Readiness Quiz

Subtract.

1.   1,025
      −   6

2.   2,642
      −   8

3.   3,467
      −   29

4. 7,024 − 15 = _____

5. 1,331 − 17 = _____

6. 6,050 − 23 = _____

7. There are 1,080 pages in Gerardo's book.
He has read 1,038 pages. How many pages
are left to read?

_____

Divide.

8. 2)‾1‾6‾

9. 3)‾9‾

10. 3)‾2‾4‾

11. 35 ÷ 5 = _____

12. 48 ÷ 8 = _____

13. 56 ÷ 7 = _____

14. Sharon has $32. She wants to buy CDs that cost $8 each.
How many can she buy?

_____

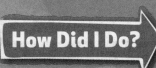

Shade the boxes to show the problems you answered correctly.

1	2	3	4	5	6	7	8	9	10	11	12	13	14

How Did I Do?

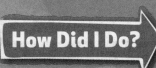

MY Math Words

Vocab
abc

Review Vocabulary

dividend   divisor   quotient

Making Connections

Read the word problem. Use the review vocabulary to describe what each number represents.

> A local pond is home to 36 Canada geese. They live in groups. Each group has 9 geese. How many Canada geese groups are there?

The total number of Canada geese is _____. This number represents the _____.

There are _____ Canada geese in each group. This number represents the _____.

The number of groups living in the pond represents the _____.

Write and solve a division sentence about the word problem. Circle the quotient.

_____

Lesson 6–2

compatible numbers

Estimate $4,588 \div 9$.

$\downarrow$

$4,500 \div 9$

compatible numbers

Lesson 6–8

partial quotients

$4 \overline{)624}$
$\underline{-500}$ $125 \leftarrow$ partial quotient
124
$\underline{-100}$ $25 \leftarrow$ partial quotient
24
$\underline{-24}$ $6 \leftarrow$ partial quotient
$125 + 25 + 6 = 156$

Lesson 6–3

remainder

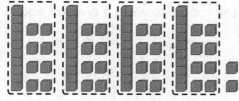

$74 \div 4 = 18 \text{ R}2$

Ideas for Use

- Work with a partner to name the part of speech of each word. Consult a dictionary to check your answers.

- Use the blank cards to write your own vocabulary cards.

A dividing method in which the dividend is separated into sections that are easy to divide.

How can the meaning of *partial* help you remember this vocabulary word?

_____

_____

Numbers in a problem that are easy to work with mentally.

How can you use basic facts to estimate a quotient?

_____

_____

The number that is left after one whole number is divided by another.

Use the model to write a division equation.

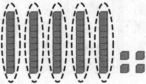

_____

MY Foldable

FOLDABLES® Follow the steps on the back to make your Foldable.

Divide

$$2\overline{)4\,2}$$

$$2\overline{)3\,7}$$ R

$$3\overline{)3\,8}$$

Multiply

Subtract Compare

Bring Down Start Over

Remainder?

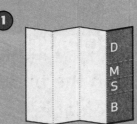

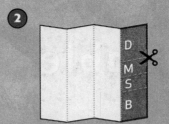

Bits

Smaller

Makes

Dividing

Number and Operations in Base Ten
4.NBT.1, 4.NBT.6, 4.OA.4
CCSS

Divide Multiples of 10, 100, and 1,000

Lesson 1

ESSENTIAL QUESTION
How does division affect numbers?

You use place value and patterns to divide dividends that are multiples of 10, 100, and 1,000.

 Math in My World Watch Tutor

Example 1

Anita's family went on vacation to an amusement park. The park has 5 entrances. 1,500 people entered the park and separated into equal lines. How many people are in each line?

Divide 1,500 people into 5 equal groups.

One Way   Use a multiplication pattern.

$5 \times 3 = 15$ $\longrightarrow$ $15 \div 5 = 3$

$5 \times 30 = 150$ $\longrightarrow$ $150 \div 5 = 30$

$5 \times 300 = 1,500$ $\longrightarrow$ $1,500 \div 5 = \underline{\hspace{2cm}}$

Another Way   Use a basic fact and place value.

$15 \div 5 = 3$ $\longleftarrow$ (basic fact)

$150 \div 5 = 30$ $\longleftarrow$ 150 is 10 times as big as 15. So, the quotient 30, is 10 times as big as 3.

$1,500 \div 5 = \underline{\hspace{2cm}}$ $\longleftarrow$ 1,500 is 100 times as big as 15. So, the quotient is 100 times as big as 3.

So, there are _____ people in each line.

Online Content at connectED.mcgraw-hill.com

Example 2

Find the quotient of 2,400 and 4.

Find $2,400 \div 4$.

One Way Use a multiplication pattern.

$4 \times 6 = 24$ ⟶ $24 \div 4 = 6$

$4 \times 60 = 240$ ⟶ $240 \div 4 = 60$

$4 \times 600 = 2,400$ ⟶ $2,400 \div 4 = $ _____

Another Way Use a basic fact and place value.

$24 \div 4 = 6$ ⟵   (basic fact)

$240 \div 4 = 60$ ⟵   ($240 = 10 \times 24$. So, $60 = 10 \times 6$.)

$2,400 \div 4 = $ _____   ⟵   ($2,400 = 100 \times 24$. So, the quotient is 100 times as big as 6.)

So, $2,400 \div 4 = $ _____.

Check

Use multiplication to check division.

$2,400 \div 4 = $ _____

_____ $\times 4 = 2,400$

Talk MATH
What basic fact will help you find the quotient of 4,200 and 7?

Guided Practice ✓

Complete each set of patterns.

1. $12 \div 4 = $ _____

$120 \div 4 = $ _____

$1,200 \div 4 = $ _____

2. $36 \div 9 = $ _____

$360 \div 9 = $ _____

$3,600 \div 9 = $ _____

Divide. Use patterns and place value.

3. $\$400 \div 2 = $ _____

4. $1,600 \div 4 = $ _____

Independent Practice

Complete each set of patterns.

5. $12 \div 2 =$ _____

$120 \div 2 =$ _____

$1,200 \div 2 =$ _____

6. $54 \div 9 =$ _____

$540 \div 9 =$ _____

$5,400 \div 9 =$ _____

7. $\$36 \div 4 =$ _____

$\$360 \div 4 =$ _____

$\$3,600 \div 4 =$ _____

8. $42 \div 6 =$ _____

$420 \div 6 =$ _____

$4,200 \div 6 =$ _____

9. $\$28 \div 7 =$ _____

$\$280 \div 7 =$ _____

$\$2,800 \div 7 =$ _____

10. $\$72 \div 8 =$ _____

$\$720 \div 8 =$ _____

$\$7,200 \div 8 =$ _____

Divide. Use patterns and place value.

11. $200 \div 5 =$ _____

12. $\$600 \div 3 =$ _____

13. $900 \div 3 =$ _____

14. $800 \div 2 =$ _____

15. $\$1,400 \div 7 =$ _____

16. $4,500 \div 5 =$ _____

17. $\$3,500 \div 5 =$ _____

18. $6,300 \div 9 =$ _____

19. $\$6,400 \div 8 =$ _____

20. $1,600 \div 8 =$ _____

21. $5,400 \div 6 =$ _____

22. $\$8,100 \div 9 =$ _____

Problem Solving

Animals migrate due to factors such as climate and food availability. The table shows a few migration distances.

Migration	
Animals	**Distance (in miles)**
Caribou	2,400
Desert locust	2,800
Green sea turtle	1,400

23. Suppose a group of green sea turtles travels 7 miles a day. How many days will the migration take?

My Work!

_____

24. Mathematical **PRACTICE** 4 **Model Math** A herd of caribou migrated the distance shown in 8 months. If they traveled the same distance each month, how many miles did the herd travel each month?

_____

HOT Problems

25. Mathematical **PRACTICE** 5 **Use Mental Math** Using mental math, tell which has a greater quotient, 1,500 ÷ 3 or 2,400 ÷ 6? Explain.

_____

_____

26. Mathematical **PRACTICE** 1 **Plan Your Solution** Complete the equation.

$\boxed{},80\boxed{} \div 6 = \boxed{}\boxed{}\boxed{}$

27. **Building on the Essential Question** Why are basic facts needed when dividing large numbers?

_____

_____

_____

MY Homework

Lesson 1

Divide Multiples of 10, 100, and 1,000

Homework Helper

Need help? connectED.mcgraw-hill.com

Find 2,700 ÷ 9.

The dividend, 2,700, is a multiple of 100. You can use a basic fact and place value to solve.

$27 ÷ 9 = 3$ ◄— This is the basic fact.

$270 ÷ 9 = 30$ ◄— See the pattern: 270 is 10 × 27, and 30 is 10 × 3.

$2,700 ÷ 9 = 300$ ◄— Continue the pattern: 2,700 is 100 × 27, and 300 is 100 × 3.

So, 2,700 ÷ 9 = 300.

Practice

Complete each set of patterns.

1. $24 ÷ 3 =$ _____

$240 ÷ 3 =$ _____

$2,400 ÷ 3 =$ _____

2. $32 ÷ 8 =$ _____

$320 ÷ 8 =$ _____

$3,200 ÷ 8 =$ _____

3. $45 ÷ 5 =$ _____

$450 ÷ 5 =$ _____

$4,500 ÷ 5 =$ _____

4. $56 ÷ 8 =$ _____

$560 ÷ 8 =$ _____

$5,600 ÷ 8 =$ _____

Divide. Use patterns and place value.

5. $1,000 ÷ 2 =$ _____

6. $500 ÷ 10 =$ _____

7. $300 ÷ 5 =$ _____

8. $2,100 ÷ 3 =$ _____

9. $7,200 ÷ 9 =$ _____

10. $\$2,000 ÷ 4 =$ _____

11. $4,200 ÷ 7 =$ _____

12. $\$2,400 ÷ 6 =$ _____

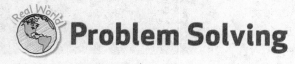
Mathematical PRACTICE ⑤ **Use Math Tools** The electronics store sold 4 laptop computers in one day. The total cost of the laptops was $3,600. If each laptop cost the same amount, how much did each laptop cost?

_____

14. The Thompson family is traveling 1,500 miles to visit relatives. They plan to drive the same distance each day. If the Thompsons make the trip in 3 days, how far will they drive each day?

_____

If they make the trip in 5 days, how far will they drive each day?

_____

15. Linus has 160 baseball cards he wants to give to his 4 cousins. If he divides the cards equally, how many cards will he give to each cousin?

_____

16. Last year, Carlotta earned $1,200 babysitting. Carlotta charges $6 per hour. What is the total number of hours Carlotta spent babysitting last year?

_____

Test Practice

17. On a trip to New York City, 8 people spent a total of $2,400 on hotel rooms. If they shared the cost equally, how much did each person spend?

Ⓐ $400   Ⓒ $40

Ⓑ $300   Ⓓ $30

Number and Operations in Base Ten
4.NBT.3, 4.NBT.6

CCSS

Estimate Quotients

Lesson 2

ESSENTIAL QUESTION
How does division affect numbers?

There are different ways to estimate quotients. One way is to use compatible numbers. **Compatible numbers** are numbers that are easy to compute mentally.

 Math in My World

Example 1

Circuses have been around for more than 200 years. They sometimes travel by train. Suppose a circus travels 642 miles in 8 hours. Estimate the quotient of 642 and 8 to find about how many miles per hour the train travels.

Estimate $642 \div 8$.

$$642 \div 8$$

642 is close to 640.

640 and 8 are compatible numbers because they are easy to divide mentally.

$$640 \div 8 = \text{\_\_\_\_}$$

 **Helpful Hint**
64 and 8 are members of a fact family.
$$8 \times 8 = 64$$
$$64 \div 8 = 8$$

So, the train is traveling about 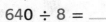 miles per hour.

Online Content at ⤳ connectED.mcgraw-hill.com

Example 2

Isabella has 6 tea sets in her collection. The collection is worth $1,168. Each tea set is worth the same amount of money. About how much is each tea set worth?

You need to estimate $1,168 ÷ 6.

One Way Use compatible numbers.

$$\$1,168 \div 6$$

$1,168 is close to $1,200. $1,200 and 6 are compatible numbers because they are easy to divide mentally.

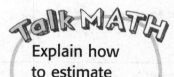

Helpful Hint

12 ÷ 6 = 2, so
1,200 ÷ 6 = 200

$$\$1,200 \div 6 = \$\underline{\hspace{1cm}}$$

Another Way Use a basic fact and place value.

$1,168 ÷ 6 ← What basic multiplication fact is close to the numbers in the problem?

$$6 \times 2 = 12$$

$$6 \times 20 = 120$$

$$6 \times \underline{\hspace{1cm}} = 1,200$$

So, each tea set is worth about _____.

Check

Use multiplication to check division.

$$1,200 \div 6 = \underline{\hspace{1cm}}$$

$$\underline{\hspace{1cm}} \times 6 = 1,200$$

Talk MATH

Explain how to estimate $4,782 ÷ 6.

Guided Practice

1. Estimate. Check your estimate using multiplicaton.

161 ÷ 4

_____ ÷ _____ = _____

Check: _____ × _____ = _____

Independent Practice

Estimate. Check your estimates using multiplication.

2. $123 \div 3$

3. $\$244 \div 6$

4. $162 \div 2$

5. $345 \div 7$

6. $538 \div 6$

7. $415 \div 6$

8. $\$1,406 \div 7$

9. $2,431 \div 8$

10. $\$2,719 \div 9$

Algebra Use mental math to find an estimate of the unknown number.

11. $4,187 \div 7 = f$

12. $\$7,160 \div c = \$800$

13. $8,052 \div 9 = t$

f is about _____

c is about _____

t is about _____

Problem Solving

Use the following information for Exercises 14 and 15.
Hut hiking involves hiking and spending the night in huts.

14. The total cost for 5 members in the Valdez family to hut hike for 6 days is $2,475. About how much does it cost for each family member?

_____

15. Ricardo needs to climb a 361-foot hill to get to the next hut. About how many yards away is he from the next hut? (*Hint:* 3 feet = 1 yard)

_____

My Work!

16. **Mathematical PRACTICE 4 Model Math** Terrence earned 806 points on 9 tests. If he earned about the same number of points on each test, about how many points did he earn on each test?

_____

17. A farm has 8 rows of beans. There are a total of 1,600 bean plants. Each row has the same number of bean plants. How many bean plants are in each row?

_____

HOT Problems

18. **Mathematical PRACTICE 1 Make a Plan** The estimated quotient of a division sentence is 200. What could the division sentence be?

_____

19. **Building on the Essential Question** How can you estimate quotients?

_____

_____

_____

MY Homework

Lesson 2

Estimate Quotients

Homework Helper

 eHelp

Need help? connectED.mcgraw-hill.com

Estimate $122 \div 3$.

Find compatible numbers, or numbers that are easy to divide mentally.
122 is close to 120. 120 and 3 are compatible numbers because they are easy to divide mentally.

Divide using the compatible numbers. $120 \div 3 = 40$

Helpful Hint
$12 \div 3 = 4$, so
$120 \div 3 = 40$

Check

Use multiplication. $3 \times 40 = 120$

So, a good estimate for $122 \div 3$ is 40.

Practice

Estimate. Check your estimates using multiplication.

1. $184 \div 9$	**2.** $\$149 \div 5$	**3.** $241 \div 8$
4. $\$422 \div 6$	**5.** $637 \div 8$	**6.** $\$3,611 \div 6$
7. $1,175 \div 4$	**8.** $5,421 \div 9$	**9.** $\$2,782 \div 7$

Algebra Use mental math to find an estimate of the unknown number.

10. $8,122 \div 9 = d$ **11.** $3,030 \div m = 600$ **12.** $4,883 \div 7 = h$

   d is about _____ m is about _____ h is about _____

Problem Solving

Estimate. Check your estimates using multiplication.

13. **Mathematical PRACTICE** ◢4◣ **Model Math** In August, 2,760 people attended concerts at Cooper Arena. About the same number of people went to see each of the 5 concerts. About how many people attended each concert?

_____

14. A mosaic at the art museum is divided into 6 sections that each contain about the same number of tiles. There are 2,889 tiles total in the mosaic. About how many tiles are in each section?

_____

15. Mr. Morgan owns an ice-cream shop. His shop made $1,380 last weekend. Mr. Morgan charges $2 for each scoop of ice cream. About how many scoops of ice cream were sold last weekend?

_____

Vocabulary Check

16. Circle the compatible numbers you could use to estimate $3,616 \div 9$.

   $3,700 \div 9$ $3,600 \div 9$ $3,620 \div 9$

Test Practice

17. Mrs. Scholl graded 632 tests during the school year. She had 3 helpers. About how many tests did each helper grade?

   Ⓐ 315 tests   Ⓒ 210 tests

   Ⓑ 310 tests   Ⓓ 200 tests

Hands On
Use Place Value to Divide

Build It Tools

Find 39 ÷ 3.

1 **Model the dividend, 39.**
Use the base-ten blocks to show 3 tens and 9 ones to show 39.

2 **Divide the tens.**
The divisor is 3. So, divide the tens into 3 equal groups.

There is _____ ten in each group.

3 **Divide the ones.**
Divide the ones into 3 equal groups.

There are _____ ones in each group.

Draw a picture to show the equal groups.

My Drawing!

There are _____ ten and _____ ones in each group.

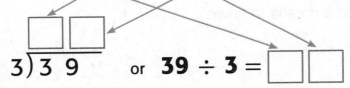

$$3{\overline{)3\ 9}} \qquad \text{or} \qquad 39 \div 3 = \boxed{}\boxed{}$$

So, the quotient is _____.

Try It

Some numbers do not divide evenly. The amount left over is called the **remainder**. Use the capital letter R to denote the remainder.

Find 68 ÷ 5 using base-ten blocks.

1 **Model the dividend.**
Use 6 tens and 8 ones to show 68.

My Drawing!

2 **Divide the tens.**
Divide the tens into 5 equal groups. There is _____ ten in each group. Regroup the ten that is left over into 10 ones.

There are _____ ones altogether.

3 **Divide the ones.**
Divide the ones into 5 equal groups. Draw a picture to show the equal groups.

There is 1 ten and 3 ones in each group. There are 3 ones left over.

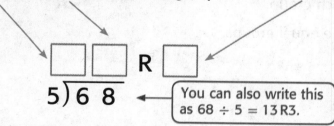

$$5\overline{)68}$$

You can also write this as 68 ÷ 5 = 13 R3.

The ones left over are the remainder.

So, 68 ÷ 5 = _____ R _____ .

Talk About It

1. **Mathematical PRACTICE 2** **Reason** Explain what it means to have a remainder when dividing.

_____

_____

Practice It

Write the division sentence shown by each model.

2.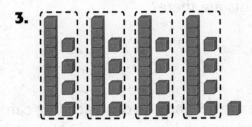

_____ ÷ _____ = _____

3.

_____ ÷ _____ = _____

Use models to find each quotient. Draw the equal groups.

4. 36 ÷ 2 = _____

5. 48 ÷ 3 = _____

There is _____ ten and _____ ones in each group.

The remainder is _____.

There is _____ ten and _____ ones in each group.

The remainder is _____.

6. 59 ÷ 4 = _____

There is _____ ten and _____ ones in each group.

The remainder is _____.

Apply It

Use models to solve.

7. There are 64 stickers. Each student gets 8 stickers. How many students are there?

_____

8. There are 73 party favors. Each bag can hold 9 party favors. How many full bags are there? How many party favors are left over?

_____

9. **Mathematical PRACTICE** 3 **Find the Error** A seal trainer has 47 seal treats. There are 4 seals. To find the number of seal treats each seal would receive, Mary drew the picture at the right to model 47 ÷ 4.

Look at Mary's drawing. Describe her error.

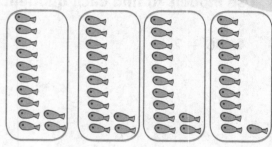

_____

_____

_____

Draw a correct picture to find 47 ÷ 4.

So, each seal will receive _____ treats.

How many seal treats are left over? _____

My Drawing!

Write About It

10. How can place value help me divide?

_____

_____

_____

MY Homework

Homework Helper

Need help? connectED.mcgraw-hill.com

Some numbers don't divide evenly. In this case, there is a remainder.

Find 43 ÷ 3.

1 Model the dividend, 43.

2 Divide the tens. The divisor is 3. So, divide the tens into 3 equal groups.

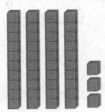

3 There is 1 ten left over. Regroup that ten as 10 ones.

4 10 ones plus the 3 ones you had to begin with makes 13 ones. Divide the 13 ones into four equal groups.

There are 1 ten and 4 ones in each group. There is 1 one left over. Left over ones are called the remainder. So, the quotient is 14 R1.

Practice

Write the division sentence shown by the model.

1.

_____

Write the division sentence shown by the model.

2.
 : _____

 Problem Solving

Use models to find each quotient. Draw the equal groups.

3. There are 70 cards. Each person gets 5 cards.
How many people are there?

$70 \div 5 =$ _____

There are _____ people.

My Drawing!

4. **Mathematical**
PRACTICE **Reason** There are 83 apples.
Each bag can hold 4 apples. How many full bags
are there? How many are left over?

$83 \div 4 =$ _____

There are _____ full bags.
There are _____ left over.

Vocabulary Check [Vocab]

5. Explain why, in division, there is sometimes a remainder.

_____

_____

Problem-Solving Investigation
STRATEGY: Make a Model

Learn the Strategy

Ann's class bought 4 boxes of peaches at the orchard. There are 128 peaches in all. Each box has the same number of peaches. How many are in each box?

I do declare!

1 Understand

What facts do you know?

There are _____ peaches divided evenly into _____ boxes.

What do you need to find?

Find the number of _____.

2 Plan

I will use base-ten blocks to model 128 ÷ 4.

3 Solve

Model 128. Divide the tens into four equal groups. Then divide the ones into four equal groups.

Since the hundred cannot be divided into 4 groups, trade 1 hundred for 10 tens.

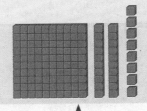

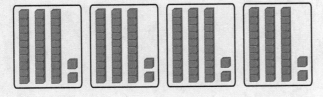

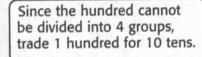

So, each box has _____ peaches.

4 Check

I can check by using repeated addition. 32 + 32 + 32 + 32 = 128

So, I know that my answer is reasonable.

Practice the Strategy

Caleb's family spent $420 on their road trip. Their road trip was 4 days long. If they spent the same amount of money each day, how much did they spend each day?

1 Understand

What facts do you know?

_____

_____

What do you need to find?

_____

2 Plan

_____

3 Solve

4 Check

Does your answer make sense? Explain.

_____

Apply the Strategy

Solve each problem by making a model.

1. Casey's mom is the baseball coach for his team. She spent $150 on baseballs. Each baseball cost $5. How many baseballs did she buy?

_____

My Work!

2. **Mathematical PRACTICE 2** **Reason** Each flowerpot costs $7. How many flowerpots can be bought with $285? Explain.

_____

_____

_____

3. Cindy spent $6 on her meal. She spent $4 on a sandwich and the rest on a smoothie. How many smoothies could be bought with $124?

_____

4. **Mathematical PRACTICE 5** **Use Math Tools** Use models to find the unknown numbers.

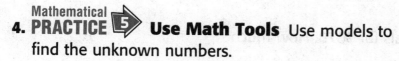

254 ÷ ☐ = ☐ R4

5. Quincy found 120 shells during four days at the beach. If Quincy found the same number of shells each day, how many shells did he find each day?

_____

Review the Strategies

Use any strategy to solve each problem.

- Make a table.
- Choose an operation.
- Act it out.
- Draw a picture.

6. The calendar shows the number of days Carlota rides her bike each month. Each time she rides her bike, she travels 10 miles. Is it reasonable to say that Carlota will bike more than 500 miles in 6 months? Explain.

September						
Sun	**Mon**	**Tues**	**Wed**	**Thurs**	**Fri**	**Sat**
					1	2 (B)
3 (B)	4	5	6	7	8 (B)	9
10 (B)	11	12	13 (B)	14	15	16 (B)
17	18	19	20	21	22 (B)	23
24 (B)	25	26	27	28 (B)	29	30 (B)

My Work!

_____

_____

_____

7. Mathematical
PRACTICE 2 **Reason** Paz and her scout troop make 325 granola bars for a fundraiser. Four granola bars are put in each bag. Paz says that there will not be any left over. Find and correct her mistake.

_____

_____

_____

8. A coach ordered 6 soccer goals for $678. How much did each goal cost?

_____

9. Gabriel has 268 miniature trains. He lines them up in 2 equal rows. How many trains are in each row?

_____

MY Homework

Homework Helper

Need help? connectED.mcgraw-hill.com

Daphne bought her mother a bunch of 12 flowers. Two of the flowers are daisies. She divided the remaining flowers in two groups. One group has tulips. How many of the flowers are tulips?

1 Understand

What do you know?

Daphne bought 12 flowers. Two are daisies. She divided the remaining flowers into two groups. One group has tulips.

What do you need to find?

I need to find how many flowers are tulips.

2 Plan

I will subtract the number of daisies. Then I will divide the remaining number of flowers by 2.

3 Solve

12 flowers − 2 daisies = 10 flowers

10 flowers ÷ 2 = 5 tulips

So, there are 5 tulips.

4 Check

I will use addition to check.

5 tulips + 5 other flowers + 2 daisies = 12 flowers

So, the answer is correct.

Problem Solving

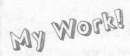

Solve each problem by making a model.

1. There are two equal groups of 8 books on the shelf. One group has reading books. In the other group, three of the books are math books, and the rest are science books. How many science books are there?

_____

2. **Mathematical**
   PRACTICE 5 **Use Math Tools** There are 364 trees planted in 2 equal rows. One row has pine trees. The rest are oak trees. How many oak trees are there?

_____

3. There are 3 types of fish in a pond. There are a total of 162 fish. There is the same number of each type of fish. How many of each type of fish are in the pond?

_____

4. Monica and her parents spent $170 on new swimsuits. Monica's swimsuit cost $35. Her mom's swimsuit cost twice as much as Monica's. How much did her dad's swimsuit cost?

_____

5. Joel divided 68 CDs evenly on 4 shelves in his bedroom. The CDs are arranged in alphabetical order, with the first shelf containing the letters A–E. If there are 13 CDs filed under A–D, how many CDs are filed under the letter E?

_____

Divide with Remainders

Lesson 5

ESSENTIAL QUESTION
How does division affect numbers?

You have used models and fact families to divide. You can also use place value.

 Math in My World Tools Watch Tutor

Example 1

Nolan and his family went to a water park during their vacation. Each seat on a water ride can hold 2 people. There are 39 people. How many seats will be needed?

Find 39 ÷ 2.

1 Divide the tens.

How many groups of 2 are in 3 tens?

_____ group of ten

2 Multiply, subtract, and compare.

Multiply. $2 \times 1 =$ _____

Subtract. $3 - 2 =$ _____

Compare. $1 < 2$

3 Bring down the ones.

Bring down 9 ones. There are now _____ ones.

4 Divide the ones.

How many groups of 2 are in 19? _____ groups

Multiply. $2 \times 9 =$ _____

Subtract. $19 - 18 =$ _____

Compare. $1 < 2$

19 seats are full. One additional seat has 1 person.

So, _____ seats are needed.

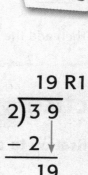

Check Use models to check.

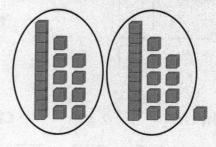

Online Content at ⟋**connectED.mcgraw-hill.com**

Example 2

Find 85 ÷ 3.

 Divide the tens. → 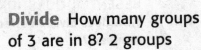 R 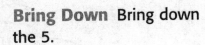 **Divide the ones.**

$3\overline{)8\ 5}$

Divide How many groups of 3 are in 8? 2 groups

Multiply $2 \times 3 = 6$

Subtract $8 - 6 = 2$

Compare. $2 < 3$

Bring Down Bring down the 5.

Divide How many groups of 3 are in 25? 8 groups

Multiply $8 \times 3 = 24$

Subtract $25 - 24 = 1$

Compare. $1 < 3$

Bring Down Since there are no numbers to bring down, 1 is the remainder.

So, 85 ÷ 3 = _____ .

Check

85 ÷ 3 = □ R □

□ × 3 = 84   Then add the remainder.   84 + □ = 85

Guided Practice ✓

Divide. Use multiplication to check.

1. 1 □ R □
$2\overline{)2\ 7}$
− □
□ 7
− □
□

2. □ □ R □
$5\overline{)5\ 9}$
− □
□ 9
− □
□

Check: □ × 2 = □
□ □ + □ = □

Check: □ × 5 = □
□ □ + □ = □

Talk MATH

When you divide a number by 6, can the remainder be 6? Explain.

Independent Practice

Divide. Use multiplication to check.

3. 4⟌48

4. 5⟌53

5. 6⟌67

6. 3⟌33

7. 7⟌73

8. 9⟌96

9. 69 ÷ 3 = _____

10. 77 ÷ 3 = _____

11. 99 ÷ 4 = _____

Algebra Use mental math to find the unknown.

12. $x \div 2 = 12$

13. $48 \div 4 = y$

14. $75 \div 5 = s$

$x =$ _____

$y =$ _____

$s =$ _____

Mathematical
15. PRACTICE 4 **Model Math** There are eight lions, four tigers, five cheetahs, six giraffes, seven hippos, and 78 monkeys at the City Zoo.

If each of the four zookeepers feeds the same number of animals, how many animals does each zookeeper feed? Explain.

_____

_____

16. Marlene makes $4 an hour babysitting. If she earned $48, how many hours did she babysit?

_____

17. Seven scouts need to sell 75 boxes of cookies. Each scout gets the same number of boxes. How many boxes will be left to sell?

_____

My Work!

HOT Problems

Mathematical
18. PRACTICE 1 **Keep Trying** Identify a two-digit dividend that will result in a quotient with a remainder of 1 when the divisor is 4.

_____

19. **Building on the Essential Question** Why is the remainder always less than the divisor? Explain.

_____

_____

_____

MY Homework

Homework Helper

Need help? ↗ connectED.mcgraw-hill.com

Find 62 ÷ 4.

 Divide the tens.

$$4\overline{)62}^{\,1}$$

Divide. How many groups of 4 are in 6 tens? 1 group of ten. Put 1 in the quotient over the tens place.

2 Multiply, subtract, and compare.

$$
\begin{array}{r}
1 \\
4\overline{)62} \\
-\ 4 \\
\hline
2
\end{array}
$$

Multiply. $4 \times 1 = 4$
Subtract. $6 - 4 = 2$
Compare. $2 < 4$

3 Bring down the ones.

$$
\begin{array}{r}
1 \\
4\overline{)62} \\
-\ 4\downarrow \\
\hline
22
\end{array}
$$

Bring down 2 ones. There are 22 ones in all.

4 Divide the ones.

$$
\begin{array}{r}
15\ \text{R2} \\
4\overline{)62} \\
-\ 4 \\
\hline
22 \\
-\ 20 \\
\hline
2
\end{array}
$$

How many groups of 4 are in 22? There are 5 ones in each group.

Put 5 in the quotient over the ones place.

Multiply. $4 \times 5 = 20$
Subtract. $22 - 20 = 2$
Compare. $2 < 4$

So, 62 ÷ 4 = 15 R2.

Practice

Divide. Check using multiplication.

1. $5\overline{)76}$

2. $2\overline{)39}$

3. $4\overline{)95}$

4. $6\overline{)86}$

5. $8\overline{)99}$

6. $3\overline{)80}$

Problem Solving

7. **PRACTICE** **2** **Reason** Mitzi has 38 coins. She divides the coins equally among herself and her 3 brothers. How many coins does each person get? Are any coins left over?

_____

8. There are 26 third graders and 32 fourth graders going on a field trip. Each van can carry 10 students.

How many vans are needed? _____

How many students will be in each van?

_____

9. In Jonah's restaurant, each table seats 4 people. Jonah has 78 napkins to place at the tables.

How many tables can Jonah set with the napkins he has? _____

How many napkins will Jonah need to complete one more table? _____

Test Practice

10. Wes bought 62 songs. He wants to put an equal number of songs on 5 different discs. How many songs will be left over?

    Ⓐ 4 songs   Ⓒ 2 songs

    Ⓑ 3 songs   Ⓓ 1 song

Interpret Remainders

Lesson 6

ESSENTIAL QUESTION
How does division affect numbers?

What should we do when there is a remainder?
We can interpret it in many ways.

 ## Math in My World

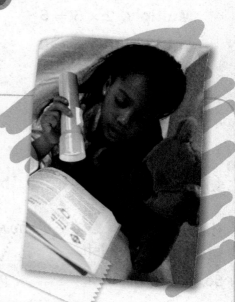

Example 1

Mandy wants to buy 4 books that each cost the same amount. If the total cost is $74, how much does each book cost?

Divide 74 by 4.

1 Divide the tens.

Divide. There is 1 group of 4 in 7.

Multiply. $4 \times 1 = 4$

Subtract. $7 - 4 = 3$

Compare. $3 < 4$

Bring Down. Bring down the ones, 4.

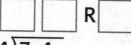
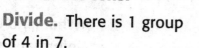

$$\begin{array}{r} \square\ \square\ R\ \square \\ 4\overline{)7\ 4} \end{array}$$

2 Divide the ones.

Divide. There are 8 groups of 4 in 34.

Multiply. $8 \times 4 = 32$

Subtract. $34 - 32 = 2$

Compare. $2 < 4$

Bring Down. Since there are no numbers to bring down, the remainder is 2.

The remainder shows that each book will cost

a little more than _____ .

Example 2

Gracie has 64 prizes. She will put 3 prizes in each bag. How many bags will she have? Interpret the remainder.

 Divide the tens.

Divide. How many groups of 3 are in 6? 2 groups

Multiply. 2 × 3 = 6

Subtract. 6 − 6 = 0

Compare. 0 < 3

Bring Down. Bring down the 4.

$$3\overline{)6\ 4}$$

R

 Divide the ones.

Divide. How many groups of 3 are in 4? 1 group

Multiply. 1 × 3 = 3

Subtract. 4 − 3 = 1

Compare. 1 < 3

Bring Down. Since there are no numbers to bring down, 1 is the remainder.

So, Gracie will have _____ bags. 64 ÷ 3 = _____ R_____

The remainder, _____, shows the number of prizes that Gracie will have left.

So, Gracie will have _____ prize left.

Check

_____ × 3 = _____ Multiply.

_____ + 1 = _____ Add the remainder.

Talk MATH

What kind of information can you get from a remainder?

Guided Practice

1. There are 45 people waiting for a bus. Each seat holds 2 people. How many seats will be needed? Divide. Interpret the remainder.

45 ÷ 2 = _____ R_____ .

So, _____ seats will be needed.

$$2\overline{)4\ 5}$$

R

5

Independent Practice

Divide. Interpret the remainder.

2. Gianna is at the school carnival. She has 58 tickets. It costs 3 tickets to play the basketball game. If she plays the basketball game as many times as she can, how many tickets will she have left?

$3\overline{)58}$

$58 \div 3 =$ _____

So, there is _____ ticket left.

3. There are 75 people waiting in line to ride a roller coaster. Each car of the roller coaster holds 6 people. How many cars will be needed?

$6\overline{)75}$

$75 \div 6 =$ _____

The answer is the next whole number, _____.

So, they will need _____ cars.

4. There are 4 cartons of orange juice in each package. If there are 79 cartons of orange juice, how many packages can be filled?

$4\overline{)79}$

$79 \div 4 =$ _____

So, _____ packages can be filled.

5. The fourth grade classes are going on a field trip. There are 90 students in all. Each van can seat 8 students. How many vans will be needed?

$8\overline{)90}$

$90 \div 8 =$ _____

The answer is the next whole number, _____.

So, they will need _____ vans.

Problem Solving

My Work!

For Exercises 6 and 7, use the following information.

Parents are driving groups of children to the science center.
Each van holds 5 children. There are 32 children in all.

Mathematical
6. PRACTICE 2 **Reason** How many vans are needed?

_____

_____

_____

7. Circle the true statement about the remainder.

- You do not need to know anything about the
  remainder to solve this problem.

- The remainder tells you that the answer is the next
  greatest whole number.

- The remainder is the answer to the question.

HOT Problems

Mathematical
8. PRACTICE 2 **Use Number Sense** Brody is organizing his
action figures on a shelf. He wants to divide them equally among
4 shelves. There are 37 action figures. Brody says he will have
2 left over. Find and correct his mistake.

_____

_____

9. ? **Building on the Essential Question** Why is it important to
know how to interpret a remainder?

_____

_____

_____

MY Homework

Lesson 6
Interpret Remainders

Homework Helper

Need help? connectED.mcgraw-hill.com

Miranda volunteers at the animal shelter. Each puppy gets 3 scoops of food per day. There are 50 scoops of food left in the bag. How many puppies will that amount feed for one day?

1 **Divide the tens.**

$$\begin{array}{r} 1 \\ 3\overline{)50} \end{array}$$

How many groups of 3 are in 5?

There is 1 ten in each group. Put 1 in the quotient over the tens place.

Multiply, subtract, and compare.

$$\begin{array}{r} 1 \\ 3\overline{)50} \\ -3\downarrow \\ \hline 20 \end{array}$$

Multiply. $3 \times 1 = 3$

Subtract. $5 - 3 = 2$

Compare. $2 < 3$, so bring down the ones to make 20.

2 **Divide the ones.**

$$\begin{array}{r} 16\ R2 \\ 3\overline{)50} \\ -3 \\ \hline 20 \\ -18 \\ \hline 2 \end{array}$$

How many groups of 3 are in 20?

There are 6 ones in each group. Put 6 in the quotient over the ones place.

Multiply. $3 \times 6 = 18$

Subtract. $20 - 18 = 2$

Compare. $2 < 3$

There are no more numbers to bring down. The remainder is 2.

So, 50 scoops of food will feed 16 puppies for one day.

The remainder shows that there will be 2 scoops of food left over.

Practice

Divide. Interpret the remainder.

1. Tyler is planting 60 trees at the apple orchard. He will plant 8 trees in each row. How many full rows of trees will Tyler plant?

   8)60

   _____ rows

2. Ms. Ling bought party hats for the 86 fourth graders at her school. The hats come in packages of 6. How many packages did Ms. Ling buy?

   6)86

   _____ packages

Problem Solving

3. Wezi has sixty-eight $1-bills. He takes the $1-bills to the bank to exchange them for $5-bills. How many $5-bills does Wezi get?

   _____

4. **Mathematical PRACTICE 2 Reason** Henry decorates the top of each cupcake with 3 walnuts. If he has 56 walnuts, is that enough to decorate 2 dozen cupcakes? Explain.

   _____

   _____

Test Practice

5. Janice bought juice packets for the 15 players on the soccer team. The juice packets come in boxes of 6. How many boxes did Janice buy?

   Ⓐ 5 boxes   Ⓒ 3 boxes

   Ⓑ 4 boxes   Ⓓ 2 boxes

Check My Progress

Vocabulary Check

For Exercises 1–2, use pictures, words, or numbers to describe each vocabulary word.

1. remainder

2. compatible numbers

Concept Check

Complete each set of patterns.

3. $36 \div 6 =$ _____

    $360 \div 6 =$ _____

    $3,600 \div 6 =$ _____

4. $21 \div 7 =$ _____

    $210 \div 7 =$ _____

    $2,100 \div 7 =$ _____

Algebra Use mental math to estimate the unknown number.

5. $2,369 \div 6 = t$

    t is about _____

6. $\$6,285 \div y = \$700$

    y is about _____

7. $4,022 \div 8 = r$

    r is about _____

Divide. Use multiplication to check.

8. $3\overline{)63}$

9. $8\overline{)43}$

10. $3\overline{)79}$

Problem Solving

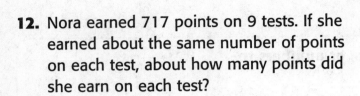

Family Vacation

Item	Total Cost
Campsite rental cost	$50
Camping supplies	$75
Food	$75

11. There are 4 members of a family planning a weekend camping trip. If the total cost is divided equally, how much will the trip cost for each person?

_____

My Work!

12. Nora earned 717 points on 9 tests. If she earned about the same number of points on each test, about how many points did she earn on each test?

_____

13. The fourth grade classes are going on a field trip. There are 85 students in all. Each van can seat 8 students. How many vans will be needed?

$85 \div 8 =$ _____

The answer is the next whole number, _____.

So, they will need _____ vans.

14. Ruby has 200 minutes left on her cell phone plan for the next 5 days. If she uses the same number of minutes each day, how many minutes can Ruby use her cell phone each day?

_____

Test Practice

15. There are 39 paintbrushes. Each art bucket holds 6 paintbrushes. How many buckets will be needed to fit all of the paintbrushes?

Ⓐ 3 buckets

Ⓑ 5 buckets

Ⓒ 7 buckets

Ⓓ 9 buckets

Number and Operations in Base Ten
4.NBT.6

CCSS

Place the First Digit

Lesson 7

ESSENTIAL QUESTION
How does division affect numbers?

Sometimes the first digit of the dividend is less than the divisor. You may not be able to place the first digit of the quotient over the first digit of the dividend.

 Math in My World

Watch ▶ Tutor 💬

Example 1

Adriana's family went to Yellowstone National Park during their vacation. Suppose one of Yellowstone's geysers erupts every 7 minutes. How many times does the geyser erupt in 65 minutes?

Find 65 ÷ 7.

Estimate **65 ÷ 7** ⟶ **70 ÷ 7 =** _____

 Divide the tens.

$7\overline{)65}$ **Divide.** 6 ÷ 7   Since 7 > 6, you cannot divide the tens.

 Divide the ones.

☐ R ☐

$7\overline{)6\ 5}$
$-6\ 3$

☐

Divide. There are 9 groups of 7 in 65.

Multiply. 7 × 9 = 63
Subtract. 65 − 63 = 2

Compare. 2 < 7
Bring Down. Since there are no more numbers to bring down, the remainder is 2.

So, the geyser will erupt 9 times in 65 minutes.

Check

9 is close to the estimate of 10. So, the answer is reasonable.

Online Content at ✐ **connectED.mcgraw-hill.com**

Example 2

A tennis coach has 125 tennis balls. There are 4 members on the team. How many balls does each player get for practice if each player gets the same number of balls?

Find 125 ÷ 4.

 Divide the hundreds.

Divide. 1 ÷ 4 Since 4 > 1, you cannot divide the hundreds.

$$\begin{array}{c} \boxed{}\,\boxed{} \text{ R } \boxed{} \\ 4\overline{)125} \\ -\boxed{} \\ \hline \boxed{}\,\boxed{} \\ -\boxed{} \\ \hline \boxed{} \end{array}$$

 Divide the tens.

Divide. 12 ÷ 4 = 3 So, write 3 in the quotient over the tens place.
Multiply. 3 × 4 = 12
Subtract. 12 − 12 = 0
Compare. 0 < 4
Bring Down. Bring down the 5.

 Divide the ones.

Divide. There is 1 group of 4 in 5.
Multiply. 1 × 4 = 4
Subtract. 5 − 4 = 1
Compare. 1 < 4
Bring Down. Since there are no more numbers to bring down, the remainder is 1.

So, each team member gets _____ balls.

There will be _____ ball left over.

Talk MATH

Estimation is one method that can be used to check division. Identify another method.

Guided Practice

Circle the correct place-value position to show where to place the first digit.

1. $2\overline{)33}$

tens

ones

2. $3\overline{)179}$

hundreds

tens

ones

Independent Practice

Divide. Use estimation to check.

3.
```
  ☐☐ R☐
2)3 7
 -☐
 ____
  ☐☐
 -☐☐
 ____
   ☐
```

4. 5)49 R

5. 6)91 R

Estimate: _____   Estimate: _____   Estimate: _____

Divide. Use multiplication to check.

6.
```
  ☐☐ R☐
4)7 9
 -☐
 ____
  ☐☐
 -☐☐
 ____
   ☐
```

7. 2)151 R

8. 3)286 R

Check: ____ × ____ = ____   Check: _____   Check: _____

____ + ____ = ____

Problem Solving

Every month, Americans throw out enough bottles and jars to fill up a giant skyscraper. All of these bottles and jars could be recycled.

My Work!

9. When one aluminum can is recycled, enough energy is saved to run a television for 3 hours. How many cans need to be recycled to run a television for 75 hours?

_____

10. **Mathematical PRACTICE 2 Reason** Most Americans use 7 trees a year in products that are made from trees. How old is a person who has used 65 trees?

_____

HOT Problems

11. **Mathematical PRACTICE 1 Plan Your Solution** When Kira's father's age, a, is divided by Kira's age, b, you get a quotient of 13 R1. Identify one possibility for their ages.

$a =$ _____ $b =$ _____

12. **Mathematical PRACTICE 3 Find the Error** Colton is finding $53 \div 3$. Find and correct his mistake.

$$
\begin{array}{r}
11 \\
3\overline{)53} \\
-3\downarrow \\
\hline
3 \\
-3 \\
\hline
0
\end{array}
$$

_____

_____

_____

13. **Building on the Essential Question** How do I know where to place the first digit of a quotient in a division problem?

_____

_____

_____

MY Homework

Homework Helper

Need help? connectED.mcgraw-hill.com

Find 145 ÷ 3.

 Divide the hundreds.

$3 \overline{)145}$

Sometimes it is not possible to divide the first digit of the dividend by the divisor. Since 3 is greater than 1, you cannot divide the hundreds.

2 Divide the tens.

$$\begin{array}{r} 4 \\ 3\overline{)145} \\ -12 \\ \hline 25 \end{array}$$

Divide. 14 ÷ 3 = 4

Multiply. 3 × 4 = 12

Subtract. 14 − 12 = 2

Compare. 2 < 3, so bring down the ones to make 25.

3 Divide the ones.

$$\begin{array}{r} 48 \text{ R1} \\ 3\overline{)145} \\ -12 \\ \hline 25 \\ -24 \\ \hline 1 \end{array}$$

Divide. 25 ÷ 3 = 8

Multiply. 3 × 8 = 24

Subtract. 25 − 24 = 1

Compare. 1 < 3

There are no more numbers to bring down; the remainder is 1.

Practice

Divide. Use estimation or multiplication to check.

1. $6\overline{)89}$ ^R

2. $2\overline{)73}$ ^R

3. $7\overline{)451}$ ^R

Divide. Use estimation or multiplication to check.

4. $3\overline{)105}$ 5. $4\overline{)219}$ R   6. $8\overline{)254}$ R

7. $7\overline{)688}$ R   8. $5\overline{)396}$ R   9. $6\overline{)372}$

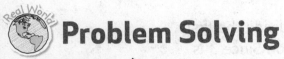

Problem Solving

10. **Mathematical PRACTICE 5 Use Math Tools** Chloe puts 4 soaps and two bottles of lotion in each gift basket. She has 127 soaps and 85 bottles of lotion. How many gift baskets can Chloe complete?

_____

11. Dante has been putting one penny in his bank every day. Now he has 161 pennies. For how many weeks has Dante been adding pennies to his bank?

_____

12. Nikki wants to buy a dog bed that costs $87. She earns $6 per week for doing chores. How many weeks will Nikki have to do her chores in order to earn enough money to pay for the dog bed?

_____

Test Practice

13. Ty's scout troop is picking up litter at the park. Each trash bag holds 8 pounds of litter. The scout troop collected a total of 230 pounds of litter. How many trash bags did they use?

Ⓐ 28   Ⓒ 30

Ⓑ 29   Ⓓ 31

Hands On
Distributive Property and Partial Quotients

Lesson 8

ESSENTIAL QUESTION
How does division affect numbers?

You have used the Distributive Property to multiply. It can also help you divide.

Draw It

Jake and his three brothers all went on separate vacations. They traveled 484 miles in all. They each traveled the same distance. How many miles did each person travel?

Find $484 \div 4$.

1 Model 484 as (400 + 80 + 4).

400	80	4

2 Divide each section by 4. Write each quotient above the bar.

4	400	80	4

$400 \div 4 = $ _____

$80 \div 4 = $ _____

$4 \div 4 = $ _____

3 Add the quotients. _____ + _____ + _____ = _____

$484 \div 4 = $ _____.

So, each person traveled _____ miles.

Check Use multiplication to check a division problem.

$484 \div 4 = $ _____

_____ $\times\ 4 = 484$

Partial quotients is a way to divide where you break the dividend into parts that are easier to divide.

Try It

There are 625 bars of soap. Each gift bag will have 5 bars of soap. How many gift bags can be made with the soaps?

Find 625 ÷ 5 using partial quotients.

 Divide the hundreds.
500 is close to 625 and is compatible with 5.
Divide 500 by 5.

_____ is a partial quotient.

Subtract 500 from 625.

Partial Quotients

```
5)625
 − 500        _____
   125
 − 100        _____
    25
 −  25        _____
     0
```

Divide the tens.
100 is close to 125 and is compatible with 5.
Divide 100 by 5.

_____ is a partial quotient.

Subtract 100 from 125.

Divide the ones.
Divide 25 by 5.

_____ is a partial quotient.

So, _____ gift bags can be made.

Add the partial quotients.

_____ + _____ + _____ = _____

625 ÷ 5 = _____

Talk About It

1. Draw an area model that could be used to find 346 ÷ 2 by using the Distributive Property.

Mathematical
2. **PRACTICE** **3** **Justify Conclusions** How are partial quotients similar to partial products?

_____

_____

_____

Practice It

Divide. Use the Distributive Property. Complete the area models.

3. 624 ÷ 2

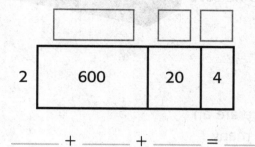

_____ + _____ + _____ = _____

624 ÷ 2 = _____

4. 848 ÷ 4

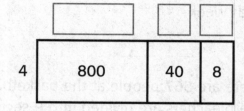

848 ÷ 4 = _____

Divide. Use the Distributive Property. Draw area models.

5. 669 ÷ 3

6. 442 ÷ 2

Divide. Use the Distributive Property or partial quotients.

7. 7)826

8. 4)924

I need a rest!

My Work!

Use the Distributive Property or partial quotients to solve Exercises 9–11.

9. Blake's dad needs $165 for a new suitcase. If he saves an equal amount for 3 weeks, how much does he save each week?

_____

10. There are 567 people at the basketball game. The bleachers are divided into 9 sections. There are an equal number of people in each section. How many people are in each section?

_____

11. **Mathematical PRACTICE** 5 **Use Math Tools** Mrs. Schmitt baked 224 cookies in her bakery. She placed them in 2 rows in her display case. How many cookies are in each row?

_____

12. In Exercise 10, why can the Distributive Property not be used to solve the problem?

_____

13. **Mathematical PRACTICE** 6 **Explain to a Friend** When finding $180 \div 4$, is 40 or 60 a more reasonable partial quotient? Explain your answer to a friend.

_____

_____

Write About It

14. Why are the Distributive Property and partial quotients helpful when dividing?

_____

_____

Name _____

MY Homework

Homework Helper

Need help? connectED.mcgraw-hill.com

Find 375 ÷ 5.

You can use the Distributive Property and an area model to divide.

1 Model 375 as (300 + 70 + 5).

300	70	5

2 Divide each section by 5.

	60	14	1
5	300	70	5

3 Add the partial quotients.

$60 + 14 + 1 = 75$

$375 \div 5 = 75$

So, $375 \div 5 = 75$.

Check

Multiply to check your answer.

$5 \times 75 = 375$, so the answer is correct.

Practice

Divide. Use the Distributive Property. Complete the area models.

1. 639 ÷ 3

_____ + _____ + _____ =

639 ÷ 3 = _____

2. 336 ÷ 6

_____ + _____ + _____ =

336 ÷ 6 = _____

Problem Solving

3. Mathematical PRACTICE 5 Use Math Tools

A garden store has 868 plants. They are divided equally into two groups. How many plants are in each group?

_____

Divide. Use the Distributive Property or partial quotients.

4.
```
  3)762
  - 600     _____
    162
  - 150     _____
     12
  -  12     _____
      0
```

_____ + _____ + _____ = _____

762 ÷ 3 = _____

5.
```
  2)426
  - 400     _____
     26
  -  20     _____
      6
  -   6     _____
      0
```

_____ + _____ + _____ = _____

426 ÷ 2 = _____

Vocabulary Check

6. Explain why it can be helpful to use partial quotients when dividing.

_____

_____

_____

Name ....

Divide Greater Numbers

Lesson 9

ESSENTIAL QUESTION
How does division affect numbers?

Dividing three- and four-digit numbers is similar to dividing two-digit numbers.

 Math in My World Watch Tutor

Example 1

There are 678 people in line to ride a roller coaster. Each coaster car holds 6 people. How many coaster cars are needed so that everyone in line rides the coaster once?

Divide 678 by 6.

1 Divide the hundreds.
Divide. $6 \div 6 = 1$
Write 1 in the hundreds place.
Multiply. $6 \times 1 = 6$
Subtract. $6 - 6 = 0$
Compare. $0 < 6$
Bring down the tens.

$$\begin{array}{r} 1 \\ 6\overline{)678} \\ -\,6 \\ \hline 07 \end{array}$$

2 Divide the tens.
Divide. There is 1 group of 6 in 7.
Write 1 in the tens place.
Multiply. $6 \times 1 = 6$
Subtract. $7 - 6 = 1$
Compare. $1 < 6$
Bring down the ones.

$$\begin{array}{r} 113 \\ 6\overline{)678} \\ -\,6 \\ \hline 07 \\ -\,6 \\ \hline 18 \\ -\,18 \\ \hline 0 \end{array}$$

3 Divide the ones.
Divide. $18 \div 6 = 3$
Write 3 in the ones place.
Multiply. $6 \times 3 = 18$
Subtract. $18 - 18 = 0$
Compare. $0 < 6$

$678 \div 6 =$ _____

So, _____ coaster cars are needed.

Example 2

A roller coaster takes about 4 minutes to travel its 1,970-foot track. How many feet does the coaster travel in one minute?

Divide 1,970 by 4.

Estimate 1,970 ÷ 4 ⟶ 2,000 ÷ 4 = _____

① **Divide the thousands.**

Since is 1 < 4, you cannot divide the thousands.

② **Divide the hundreds.**

Divide. There are 4 groups of 4 in 19.
Multiply. Subtract. Compare. Bring down.

③ **Divide the tens.**

Divide. There are 9 groups of 4 in 37.
Multiply. Subtract. Compare. Bring down.

④ **Divide the ones.**

Divide. There are 2 groups of 4 in 10.
Multiply. Subtract. Compare. Bring down.

⑤ **Find the remainder.**

```
      ☐ ☐ ☐  R☐
  4)1, 9 7 0
   - 1 6↓
       3 7
     - 3 6↓
         1 0
         - 8
           2
```

So, it travels a little more than _____ feet each minute.

Check The answer, a little more than _____, is close to the estimate of 500. So, the answer is reasonable.

Talk MATH

How would you mentally determine the number of digits in the quotient for 795 ÷ 5?

Guided Practice

Divide. Use estimation to check.

```
      ☐ ☐ ☐
 1. 2)286
    - 2↓
      08↓
    - 8↓
      06
    - 6
      0
```

2. 2)745

Independent Practice

Divide. Use estimation to check.

3. 2)324

4. 3)585

5. 2)1,573

Estimate:

Estimate:

Estimate:

Divide. Use multiplication to check.

6. 3)787

7. 2)849

8. 4)994

Check:

Check:

Check:

9. 3)1,863

10. 4)3,974

11. 4)2,611

Check:

Check:

Check:

Problem Solving

Use the following information for Exercises 12–13.

The White House is the official home and workplace of the President of the United States. President Theodore Roosevelt gave the White House its name, based on its color.

12. **Mathematical PRACTICE 2** **Reason** It takes 570 gallons of paint to paint the outside of the White House. If the number of gallons used to paint each of its 4 sides is equal, how many gallons of paint are used on each side?

_____

13. There are 132 rooms and 6 floors in the White House. If each floor has the same number of rooms, how many rooms would each floor have?

_____

14. Britney reads a book in 9 days. If the book is 1,116 pages long, and she reads the same number of pages each day, how many pages does she read each day?

_____

HOT Problems

15. **Mathematical PRACTICE 1** **Make a Plan** Write a division problem that results in a quotient that is greater than 200 and less than 250.

_____

16. **Building on the Essential Question** Do the quotients always have the same number of digits when dividing 3-digit numbers by 1-digit numbers?

_____

_____

_____

MY Homework

Homework Helper

Need help? connectED.mcgraw-hill.com

Find 1,927 ÷ 4.

Estimate 1,927 is close to 2,000. 2,000 ÷ 4 = 500.

1 Divide the thousands.

$$4\overline{)1,927}$$

Sometimes it is not possible to divide the first digit of the dividend by the divisor.
Since 4 > 1, you cannot divide the thousands.

2 Divide the hundreds.

$$\begin{array}{r} 4 \\ 4\overline{)1,927} \\ -16 \\ \hline 32 \end{array}$$

Divide. 19 ÷ 4 = 4
Multiply. 4 × 4 = 16
Subtract. 19 − 16 = 3
Compare. 3 < 4
Bring down the 2.

3 Divide the tens.

$$\begin{array}{r} 48 \\ 4\overline{)1,927} \\ -16 \\ \hline 32 \\ -32 \\ \hline 07 \end{array}$$

Divide. 32 ÷ 4 = 8
Multiply. 4 × 8 = 32
Subtract. 32 − 32 = 0
Compare. 0 < 4
Bring down the 7.

4 Divide the ones.

$$\begin{array}{r} 481\text{R}3 \\ 4\overline{)1,927} \\ -16 \\ \hline 32 \\ -32 \\ \hline 07 \\ -4 \\ \hline 3 \end{array}$$

Divide. There is 1 group of 4 in 7.
Multiply. 4 × 1 = 4
Subtract. 7 − 4 = 3
Compare. 3 < 4
Since there are no more numbers to bring down, the remainder is 3.

1,927 ÷ 4 = 481 R3

Check

481 R3 is close to the estimate, 500, so the answer is reasonable.

Practice

Divide. Use estimation or multiplication to check.

1. 3)534

2. 7)2,761

3. 4)850

4. 8)1,074

5. 5)3,344

6. 6)5,244

 **Problem Solving**

7. Ann needs to read 414 pages in 3 days. How many pages should she read each day?

_____

8. **Mathematical PRACTICE 5** **Use Math Tools** Marcel received an award for community service. The award included a check for $2,265. Three businesses each contributed the same amount toward the prize money. How much did each business contribute?

_____

Test Practice

9. Eric collected 560 books to donate to a preschool. The books were divided equally among 5 classrooms. How many books did each classroom get?

   Ⓐ 112 books   Ⓒ 132 books

   Ⓑ 110 books   Ⓓ 512 books

Vocabulary Check

1. Circle which of the following uses **partial quotients** to divide 362 by 2.

```
    181          181
  2)362        2)362      100
  − 2↓         − 200
   16           162        80
  − 16↓        − 160
    02            2       + 1
   − 2                    181
    0
```

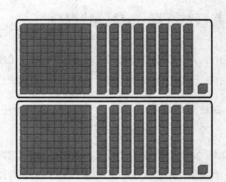

Concept Check

Divide. Use estimation to check.

2. 4)56

3. 5)71 R

Estimate:

_____ ÷ _____ = _____

Estimate:

_____ ÷ _____ = _____

Divide. Use multiplication to check.

4. $3\overline{)345}$ 5. $3\overline{)679}$

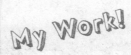

6. $697 \div 7 =$ _____   7. $883 \div 9 =$ _____

8. $917 \div 4 =$ _____   9. $775 \div 5 =$ _____

Problem Solving

10. There are 78 campers at a summer camp. There are 6 campers per cabin. How many cabins are there? _____

11. Carlo has $46 to spend on trading cards. If each pack of cards costs $3, how many packs of cards can he buy? _____

12. A tug-of-war team weighs a total of 774 pounds. The 6 members on the team each weigh the same amount. How much does each person weigh? _____

13. A coach ordered 9 hockey goals for $4,050. How much did each goal cost? _____

Test Practice

14. There are 456 runners in a race. There are 4 groups of runners. Each group has the same number of runners. How many runners are in each group?

   Ⓐ 111 runners   Ⓒ 113 runners

   Ⓑ 112 runners   Ⓓ 114 runners

Quotients with Zeros

Lesson 10

ESSENTIAL QUESTION
How does division affect numbers?

In division, a quotient will sometimes contain zeros.

I'm ready for my close-up!

 ## Math in My World

Watch ▶ Tutor 💬

Example 1

The Ramos family is going on a vacation to the behind-the-scenes tour of a wildlife reserve in a park. How much will it cost for each family member to go on the tour?

Cost of Tour	
Number of People	Cost ($)
3	327

Find $327 ÷ 3.

 Divide the hundreds.
3 ÷ 3 = 1
Multiply. Subtract. Compare. Bring down.

 Divide the tens.
2 < 3
There are not enough tens to divide.
Write 0 in the tens place.
Multiply. Subtract. Compare. Bring down.

 Divide the ones.
27 ÷ 3 = 9
Multiply. Subtract. Compare.

327 ÷ 3 = _____.

$$3)\overline{3\quad 2\quad 7}$$

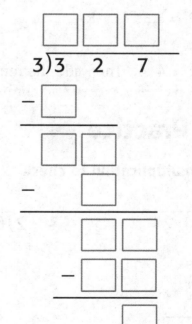

So, it will cost each family member $_____.

Example 2

The Kincaid family is going on vacation. They have to drive 415 miles to get to and from Dolphin Cove. How far is it to Dolphin Cove?

Divide 415 by 2.

 Divide the hundreds.
$4 \div 2 = 2$
Multiply. Subtract. Compare. Bring down.

 Divide the tens.
$1 < 2$. There are not enough tens to divide.
Write 0 in the tens place.
Multiply. Subtract. Compare. Bring down.

 Divide the ones.
There are 7 groups of 2 in 15.
Multiply. Subtract. Compare.

 Find the remainder.
A remainder of 1 tells you that the quotient is just over 207.

So, the distance to Dolphin Cove is a little more than _____ miles.

Check Use multiplication to check a division problem.

$415 \div 2 =$ _____ R _____

_____ $\times 2 = 414$ Then add the remainder. $414 +$ _____ $= 415$.

$$\begin{array}{r} \square\,\square\,\square \quad \text{R}\,\square \\ 2\overline{)4\quad 1\quad 5} \\ -\square \\ \hline \square\quad\square \\ -\square \\ \hline \square\quad\square \\ -\square\quad\square \\ \hline \square \end{array}$$

Guided Practice

Divide. Use multiplication to check.

1.
$$\begin{array}{r} 2\overline{)2\ 1\ 2} \\ -2 \\ \hline 0\ 1 \\ -\ 0 \\ \hline 1\ 2 \\ -\ 1\ 2 \\ \hline 0 \end{array}$$

2. $2\overline{)6\ 1\ 7}$

Talk MATH
Explain how to find the quotient of $624 \div 3$.

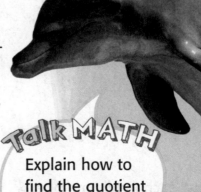

Independent Practice

Divide. Use multiplication to check.

3. 2)214

4. 3)327

5. 5)$545

6. $613 ÷ 3 = _____

7. 837 ÷ 4 = _____

8. 1,819 ÷ 2 = _____

Algebra Find the unknown.

9. 416 ÷ ■ = 208

10. 622 ÷ 3 = 207 R ■

11. $2,429 ÷ 3 = $ ■ R2

■ = _____

■ = _____

■ = _____

Problem Solving

Use the following information for Exercises 12 and 13.

Geocaching is an outdoor treasure hunting game in which participants use a Global Positioning System (GPS) to hide and seek "treasures" all over the world. The "treasures" are usually toys or trinkets.

12. Chad is saving his money to buy a GPS receiver so that he can go geocaching. He has 2 months to save $215. How much money does he need to save each month?

_____

13. **Mathematical PRACTICE 2 Reason** Some of the treasures have been hidden on mountains. If the treasure is 325 feet away, how many yards away is it? (*Hint:* 3 feet = 1 yard)

_____

14. There are 408 students at a school. There are 4 lunch periods. If there is the same number of students in each lunch period, how many students are in each period?

_____

My Work!

HOT Problems

15. **Mathematical PRACTICE 1 Keep Trying** Identify a 3-digit dividend that will result in a 3-digit quotient that has a zero in the tens place when the divisor is 6.

_____

16. **Building on the Essential Question** Why do I sometimes have to use 0 in a quotient?

_____

_____

MY Homework

Homework Helper

Need help? connectED.mcgraw-hill.com

Find 614 ÷ 3.

 Divide the hundreds.
$6 \div 3 = 2$.
Multiply. Subtract. Compare. Bring down.

 Divide the tens.
$1 < 3$ There are not enough tens to divide.
Write 0 in the tens place.
Multiply. Subtract. Compare. Bring down.

 Divide the ones.
There are 4 groups of 3 in 14.
Multiply. Subtract. Compare.

Find the remainder.

So, 614 ÷ 3 = 204 R2.

$$
\begin{array}{r}
204 \ \text{R2} \\
3\overline{)614} \\
-6 \\
\hline
01 \\
-0 \\
\hline
14 \\
-12 \\
\hline
2
\end{array}
$$

Practice

Divide.

1. 5)535

2. 4)826

3. 2)819

Divide.

4. 6)$1,824

5. 7)3,517

6. 4)2,425

7. 3)626

8. 5)$4,015

9. 8)1,613

 Problem Solving

10. The water park had a total of 1,212 visitors on Friday, Saturday, and Sunday. If the same number of people visited each day, how many visitors were there on Sunday?

_____

11. **Mathematical PRACTICE ⑤ Use Math Tools** The camping club spent $420 on four new tents. If each tent cost the same amount, how much did each tent cost?

_____

12. Mallory has 535 E-mail messages in her Inbox. She sorts them evenly among folders labeled Family, Friends, Work, School, and Recipes. How many messages does Mallory put in each folder?

_____

Test Practice

13. Mr. Lopez has collected 1,425 stamps. He sorts them into 7 piles. Which answer shows how many stamps are in each pile and how many are left over?

Ⓐ 204 R3

Ⓒ 220 R4

Ⓑ 229 R2

Ⓓ 203 R4

Name _____

Number and Operations in Base Ten
4.OA.3

Solve Multi-Step Word Problems

Lesson 11

ESSENTIAL QUESTION
How does division affect numbers?

Some word problems require more than one operation to solve.

Math in My World

 Watch   Tutor

Example 1

Jada's family bought memberships to the recreation center for a year. It cost $532 for 4 regular memberships. A deluxe membership costs an additional $35 per membership. How much does a deluxe membership cost?

Write an equation.

cost of each regular membership   add $35   deluxe membership

($532 ÷ 4) + $35 = ■ ◀── unknown

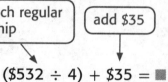

1 Divide.

$$4\overline{)5\ 3\ 2}$$

2 Add 35 to the quotient.

_____ + 35 = _____

The unknown is $ _____ .

Helpful Hint

Parentheses tell you which operations to perform first.

So, a deluxe membership costs $ _____ .

Online Content at connectED.mcgraw-hill.com   Lesson 11   393

Copyright © The McGraw-Hill Companies, Inc.   Lars A. Niki/The McGraw-Hill Companies

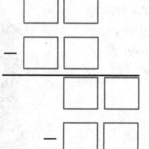

Example 2

Devin's family went snowboarding. Tickets cost $25 for kids and $30 for adults. Helmets cost $10 each. There are 3 kids and 2 adults in Devin's family. Everyone in Devin's family is getting a ticket and a helmet. Suppose they pay with $200. How much change will they receive?

Write an equation. The variable c can be used to represent the unknown.

 Find the total cost, c.

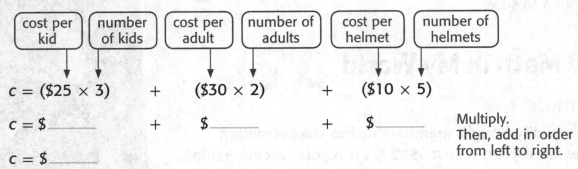

| cost per kid | number of kids | cost per adult | number of adults | cost per helmet | number of helmets |

$c = (\$25 \times 3) \quad + \quad (\$30 \times 2) \quad + \quad (\$10 \times 5)$

$c = \$\underline{\hspace{1cm}} \quad + \quad \$\underline{\hspace{1cm}} \quad + \quad \$\underline{\hspace{1cm}}$

Multiply. Then, add in order from left to right.

$c = \$\underline{\hspace{1cm}}$

The total cost is $\$\underline{\hspace{1cm}}$.

 Find the amount of change.

$\$200 - \$185 = \$\underline{\hspace{1cm}}$.

So, Devin's family received $\$\underline{\hspace{1cm}}$ in change.

Talk MATH

What kinds of words help you decide which operations to use?

Guided Practice ✓ Check

1. The regular bookstore sold 345 books. The discount bookstore sold 3 times as many books. How many books were sold altogether? Write an equation to solve the problem. Use a variable for the unknown.

$345 + (3 \times 345) = b$

$345 + \underline{\hspace{1cm}} = b$

$\underline{\hspace{1cm}} = b$

There are _____ books.

Independent Practice

**Write an equation to solve each problem.
Use a variable for the unknown.**

2. Ashlyn and her friends are making gingerbread houses. They need to divide 28 packages of gumdrops between 7 friends. There are 25 gumdrops in each package. How many gumdrops will each friend get?

Each friend gets _____ gumdrops.

3. Dominic ordered 210 pens. He divided them equally among his 10 friends. One of his friends, Benjamin, already had 27 pens. Then, Benjamin gave 13 of his pens to Peyton. How many pens does Benjamin have?

Benjamin has _____ pens.

4. Shannon is collecting art supplies. She has 48 crayons, 24 markers, and 16 stickers. She divided the crayons into 8 equal groups, the markers into 6 equal groups, and the stickers into 4 equal groups. She promised her brother that he could have one group from each type of art supplies. How many art supplies does her brother get?

Shannon's brother gets _____ art supplies.

Problem Solving

Use the following information to solve Exercises 5 and 6.

5. **Mathematical PRACTICE** 4 **Model Math** Hunter bought three boxes of cereal bars. Each box has 35 strawberry cereal bars. He also bought 20 apple cereal bars and some blueberry cereal bars. There are a total of 150 cereal bars.

Write an equation to describe the number of cereal bars that Hunter bought. How many blueberry cereal bars did Hunter buy?

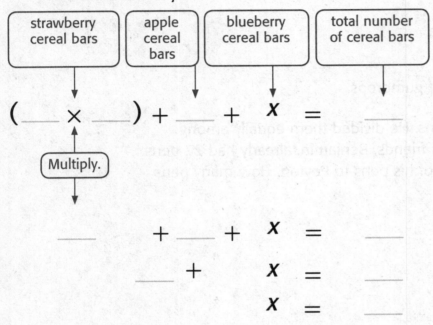

| strawberry cereal bars | apple cereal bars | blueberry cereal bars | total number of cereal bars |

(___ × ___) + ___ + **x** = ___

Multiply.

___ + ___ + **x** = ___

___ + **x** = ___

x = ___

So, there are ___ blueberry cereal bars.

HOT Problems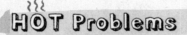

6. **Mathematical PRACTICE** 2 **Reason** Hunter wants to divide 150 cereal bars evenly among 7 friends. How many cereal bars will Hunter have left for himself?

_____

7. ? **Building on the Essential Question** How can writing equations help me solve multi-step problems?

_____

_____

MY Homework

Homework Helper eHelp

Need help? connectED.mcgraw-hill.com

Corinne has $45 saved from her allowance. She earns $7 per week. Corinne spends $2 of each allowance on a treat and saves the rest. How many weeks has Corinne been saving her money?

You need to find how much Corinne saves per week and divide $45 by that number.

1 Subtract.

$7	total weekly allowance
− $2	amount spent each week
$5	amount saved each week

2 Divide.

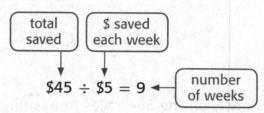

So, Corinne has been saving for 9 weeks.

Practice

1. Zoe needs to buy beads to make jewelry. For each earring, Zoe uses 3 small blue beads, 1 small metallic bead, and 2 large green beads. What is the total number of beads Zoe will need to buy to make 8 pairs of earrings?

_____

My Work!

Problem Solving

Mathematical
2. PRACTICE **2** **Reason** Miles has 327 action figures and 4 shoeboxes to store them in. Each shoebox holds 80 action figures. Does Miles have enough shoeboxes? If not, how many action figures are left over?

_____

_____

3. Kyle has a notebook for each of his 5 classes. He puts 6 stickers on each notebook. There are 10 stickers on each sheet. How many sheets of stickers will Kyle use? Write an equation to solve. Use a variable for the unknown.

_____

4. Rory is driving 584 miles to a family reunion. She will drive 300 miles by herself the first day. The second day, Rory and her cousin will share the driving equally. How many miles will her cousin drive? Write an equation to solve. Use a variable for the unknown.

_____

Test Practice

5. There are 278 houses in Hannah's neighborhood. She collected a total of $780 from her neighbors to donate to a local charity. Hannah received $5 at each house she went to. From how many houses did Hannah *not* collect money?

Ⓐ 156 houses

Ⓑ 122 houses

Ⓒ 55 houses

Ⓓ 125 houses

My Work!

Name _____

Mathematical PRACTICE 6

Multiply.

1. 429
       × 5

2. 357
       × 4

3. 189
       × 6

4. 672
       × 7

5. 2,416
       ×   3

6. 7,515
       ×   4

7. 4,219
       ×   6

8. 5,413
       ×   8

9. 3,035
       ×   2

10. 8,107
        ×   6

11. 4,050
        ×   9

12. 8,063
        ×   5

13. 83
        × 24

14. 27
        × 55

15. 64
        × 52

16. 92
        × 29

Online Content at connectED.mcgraw-hill.com

Divide.

1. 3)162

2. 5)261

3. 6)759

4. 4)529

5. 5)483

6. 4)244

7. 2)921

8. 8)327

9. 2)3,216

10. 6)4,842

11. 3)2,093

12. 5)3,526

13. 9)2,631

14. 3)5,111

15. 6)2,052

16. 4)1,729

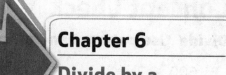
Vocabulary Check

Use the word bank to match each example with its vocabulary word.

compatible numbers	equation	partial quotients
remainder	variable	

1. $122 \div 6 \longrightarrow 120 \div 6 = 20$

_____

2.
```
      23 R4
  6)142
   −12
     22
   − 18
      4
```
↑

_____

3. $80 \div 5 = 15$

_____

4. $360 \div 9 = x$

↑

_____

5. _____

```
     125 R4
  6)754
   −600       100
    154
   −120        20
     34
   − 30       + 5
      4       125
```

Concept Check ☑

Divide. Use patterns and place value.

6. $600 \div 2 =$ _____

7. $7{,}200 \div 9 =$ _____

8. $6{,}400 \div 8 =$ _____

Estimate.

9. $715 \div 8$

10. $2{,}660 \div 9$

11. $8{,}099 \div 9$

Use the Distributive Property or partial quotients to divide.

12. $448 \div 2 =$ _____

13. $200 \div 8 =$ _____

Divide. Use multiplication to check.

14. $2\overline{)64}$

15. $7\overline{)694}$

16. $8\overline{)783}$

17. $2\overline{)2{,}157}$

18. $8\overline{)487}$

19. $3\overline{)451}$

Problem Solving

20. Clara's total score for 3 games of bowling is 312. If Clara earned the same score for each game, what was her score for each game?

_____

21. There are 3,250 buttons. If they are divided into groups of 8, how many groups are there? Interpret the remainder.

_____

_____

22. The Nair family collected 2,400 pennies. The pennies will be divided evenly among the 4 children. How many dollars will each child get?

_____

_____

23. On Saturday, 1,164 people saw a movie at Upcity Theater. There were a total of 4 movie screens with the same number of people in each audience. About how many people watched each screen?

_____

24. Holden and Alma earned $32 by doing yard work in their neighborhood. They will share their money equally. How much money will each person get?

_____

_____

Test Practice

25. Casey and her two friends made $54 selling lemonade in their neighborhood. They will share their money equally. How much money will each person get?

Ⓐ $18   Ⓒ $27

Ⓑ $20   Ⓓ $54

Reflect

Use what you learned about division to
complete the graphic organizer.

ESSENTIAL QUESTION

How does division
affect numbers?

Examples

Reflect on the ESSENTIAL QUESTION **Write your answer below.**

_____

_____

_____

Patterns and Sequences

Patterns In Our World

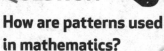

Watch a video!

MY Common Core State Standards

 CCSS

Operations and Algebraic Thinking

4.OA.3 Solve multistep word problems posed with whole numbers and having whole-number answers using the four operations, including problems in which remainders must be interpreted. Represent these problems using equations with a letter standing for the unknown quantity. Assess the reasonableness of answers using mental computation and estimation strategies including rounding.

4.OA.5 Generate a number or shape pattern that follows a given rule. Identify apparent features of the pattern that were not explicit in the rule itself.

Cool! This is what I'm going to be doing!

Standards for
Mathematical
PRACTICE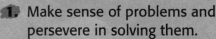

1. Make sense of problems and persevere in solving them.
2. Reason abstractly and quantitatively.
3. Construct viable arguments and critique the reasoning of others.
4. Model with mathematics.
5. Use appropriate tools strategically.
6. Attend to precision.
7. Look for and make use of structure.
8. Look for and express regularity in repeated reasoning.

 = focused on in this chapter

Thomas Tolstrup/Taxi/Getty Images   McGraw-Hill Companies, Inc.

Name _____

Am I Ready?

Check ✓ ← Go online to take the Readiness Quiz

Find each unknown.

1. 8 + _____ = 11

2. _____ + 5 = 9

3. 6 + _____ = 15

4. 13 − _____ = 7

5. _____ − 4 = 8

6. 18 − _____ = 16

7. Use the number sentence 12 + 15 + ■ = 36 to find how many books Tony read in August.

_____

Summer Reading Club	
Month	**Number of Books Read**
June	12
July	15
August	■

Find each value.

8. 8 + 1 + 6

_____

9. 7 + 2 − 3

_____

10. 2 + 10 − 6

_____

11. 11 + 6 − 6

_____

12. 12 − 3 + 4

_____

13. 16 + 4 − 10

_____

14. Each baseball uniform needs 3 buttons. Complete the table to find how many buttons are needed for 12 uniforms.

Uniforms	3	6	9	12
Buttons	9	18	27	

_____

Shade the boxes to show the problems you answered correctly.

1	2	3	4	5	6	7	8	9	10	11	12	13	14

How Did I Do?

MY Math Words

Vocab abc

Review Vocabulary

equation   operations   unknown

Making Connections

Use the review vocabulary to describe each set of examples.
Then answer the question.

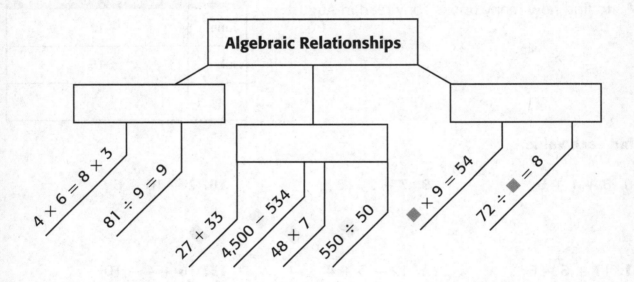

Algebraic Relationships

$4 \times 6 = 8 \times 3$

$81 \div 9 = 9$

$27 + 33$

$4,500 - 534$

48×7

$550 \div 50$

$\blacksquare \times 9 = 54$

$72 \div \blacksquare = 8$

How did you use patterns to categorize each set of examples?

_____

_____

_____

_____

Lesson 7–5

input

Input (x)	Output (y)
2	9
4	11
6	13
8	15

$$x + 7 = y$$

Lesson 7–1

nonnumeric pattern

Lesson 7–2

numeric pattern

2, 4, 8, 16, 32, 64, 128

Lesson 7–5

output

Input (x)	Output (y)
2	9
4	11
6	13
8	15

$$x + 7 = y$$

Lesson 7–1

pattern

Lesson 7–2

rule

Rule: Multiply by 4	
Number of Squares	Number of Toothpicks
1	4
2	8
3	12

Lesson 7–3

sequence

2, 4, 8, 16, 32, 64, 128

sequence

Lesson 7–3

term

2, 4, 8, 16, 32, 64, 128

term

Ideas for Use

- Arrange cards in pairs. Explain your pairings to a friend.

- Write a tally mark on each card every time you read the word in this chapter or use it in your writing. Challenge yourself to use at least 5 tally marks for each card.

Patterns that do not use numbers.

Numeric means "of or relating to numbers." How does the prefix *non*-change the meaning of *numeric*?

_____

_____

A quantity that is changed to produce an output.

What is an antonym, or opposite-meaning word, for *input*?

_____

_____

The result of an input quantity being changed.

How would you explain the difference between *input* and *output* to another student?

_____

_____

Patterns that use numbers.

Write a riddle to describe a numeric pattern. Challenge a friend to solve it.

_____

_____

_____

A statement that describes a relationship between numbers or objects.

Rule is a multiple-meaning word. Write a sentence using *rule* as a verb.

_____

_____

A list of numbers, figures, or symbols that follow a rule.

Describe a pattern you have seen on a shirt. How is this pattern similar to the math-related meaning of *pattern*?

_____

_____

_____

Each number in a numeric pattern.

Term is a multiple-meaning word. Write a sentence using another meaning of *term*.

_____

_____

The ordered arrangement of terms that make up a pattern.

Explain the word *sequence* in your own words.

_____

_____

MY Foldable

FOLDABLES® Follow the steps on the back to make your Foldable.

Input	Add 10	Subtract 3	Multiply by 3	Output Divide by 2
5	15	12	36	18
7	17	14	42	___
9	19	___		
21	31	28		
33	___	___		
___	___			

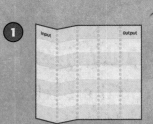

Output

Input

Nonnumeric Patterns

Lesson 1

ESSENTIAL QUESTION
How are patterns used
in mathematics?

A **pattern** is a list of numbers, figures, or symbols that follow a rule.

In this lesson, you will use and describe patterns that do not use numbers. These are called **nonnumeric patterns**. Many nonnumeric patterns are shape patterns.

Math in My World

Example 1

Edgar's room has a border of stars and moons. The stars and moons show a repeating pattern. How many figures are in this pattern unit? Copy and extend the pattern one time.

 Find the pattern unit.

1 star, 2 moons ◄ pattern unit

Each pattern unit has _____ star and _____ moons. The pattern repeats.

 Extend the pattern.
Copy the pattern as shown above. Then draw another star and two moons.

So, there are _____ figures in this pattern unit.

Patterns can also grow. They can get larger or smaller.

Example 2 Tutor

Sal uses billiard balls to show a growing pattern. Use counters to model and describe the pattern. Then make an observation about the pattern.

Use counters to model the pattern.
Begin with one counter.

Each row adds one more billiard ball. So, place _____

counters in the second row and _____ counters in the third row.

Continue adding counters until you have 5 rows.

The pattern is add _____.

The number of billiard balls in each row

alternates between odd and _____.

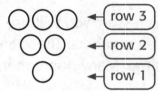
← row 3
← row 2
← row 1

Row	1	2	3	4	5
Number of Billiard Balls	1	2	3	4	5

Guided Practice Check ✓

Extend each pattern. Draw the shapes on the lines.

1. _____

2. _____

Talk MATH

Make another observation about the pattern in Example 2.

Name _____

Independent Practice

Extend each pattern. Draw the shapes on the lines.

3.  _____

4. _____ _____ _____

5. _____ _____

6. _____ _____ _____

Draw eggs in the last carton to extend each pattern.

7.

8.

Problem Solving

9. **Mathematical PRACTICE 8** **Look for a Pattern** Daniela notices a pattern in her scrapbook. The leaves on the first page are green. The second page has yellow leaves. The third page has green leaves. The next page has yellow leaves. Is this a repeating or a growing pattern? Explain.

_____

_____

_____

HOT Problems

10. **Mathematical PRACTICE 3** **Which One Doesn't Belong?**
A growing pattern is shown at the right. Which figure below would not be in this pattern? Explain.

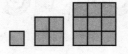

Figure 1	Figure 2	Figure 3

_____

_____

11. **?** **Building on the Essential Question** Explain the difference between a nonnumeric growing pattern and a nonnumeric repeating pattern.

_____

_____

MY Homework

Homework Helper

Need help?  connectED.mcgraw-hill.com

Extend the pattern.

 Find the pattern unit.
Each pattern unit has 1 blue pin, 1 red pin, and 1 yellow pin. This is a repeating pattern with 3 colors.

Extend the pattern.
Copy the pattern as shown above. Then draw another yellow pin.

Practice

Extend each pattern. Draw the shapes on the lines.

1.

2. ○   ⧉   ⧉⧉⧉ _____

3. ♡ ▯ ♡ ⊞ ♡ ⊞ ♡ _____

Problem Solving

Identify or extend each pattern.

4. Mathematical PRACTICE 4 Model Math
Mrs. Arthur's fourth-grade class has art every Monday. They go to computer class every Tuesday and Thursday. Mrs. Arthur's class has music on Wednesday and gym on Friday. Draw a picture to show their schedule for 2 weeks.

_____

5. Mary Anne is planting flowers in her garden. She plants tulips in the first row. She plants roses in the second row and daisies in the third row. She repeats this pattern 4 times. Which kind of flower does she plant in the tenth row? Draw a picture to show Mary Anne's garden.

_____

My Drawing!

Vocabulary Check

Write a vocabulary term to complete each sentence.

nonnumeric patterns   patterns

6. _____ are lists of numbers, figures, or symbols that follow a rule.

7. Figures or symbols presented in repeating or growing designs are
_____.

Test Practice

8. Which comes next in the pattern?

Ⓐ ◯   Ⓑ ⊗   Ⓒ ⊕   Ⓓ ⊘

Numeric Patterns

Numeric patterns are patterns that use numbers.

A pattern is a list of numbers, figures, or symbols that follows a **rule**. A rule is a statement that describes a relationship between numbers or objects. You can find and extend a pattern by following a rule.

 ## Math in My World

Example 1

Carla sells 3 picture frames for $15 and 4 picture frames for $20. The pattern for the price of the frames is the same. How much money will Carla make if she sells 6 frames?

Identify and describe the pattern by dividing the total cost by the number of frames.

Carla sells 3 picture frames for $15, and 4 picture frames for $20.

$$\$15 \div 3 = \$5$$

$$\$20 \div 4 = \$5$$

So, one picture frame costs $5. The rule is to multiply the number of frames by $5.

Use the rule to extend the pattern.

Number of Frames	1	2	3	4	5	6
Cost	$5	$10	$15	$20	$25	

+$5   +$5   +$5   +$5   +$5

So, Carla will make 6 × $5, or _____, if she sells 6 picture frames.

Example 2 Tutor

The chapters in Daniel's book follow a pattern. The even chapters have 12 pages. The odd chapters have 16 pages. The first page number of Chapter 1 is 1. Extend the pattern to find the page number of the last page in Chapter 6.

Daniel's Book	
Chapter	**Ending**
1	16
2	28
3	44
4	
5	
6	

+12
+16
+12
+16
+12

Extend the pattern.

$44 + 12 = $ ☐

☐ $+ 16 = $ ☐

☐ $+ 12 = $ ☐

So, the last page number of Daniel's book is _____.

Describe another pattern you see in this chart.

_____

_____

Talk MATH

Describe a real-world example of a growing numeric pattern.

Guided Practice ✓ Check

Identify, describe, and extend each pattern.

1. 9, 12, 15, 18, 21, _____

The pattern is add _____.

2. 5, 6, 4, 5, 3, _____

The pattern is add _____, then subtract _____.

3. Antonio runs 30 minutes each day. Extend the pattern to find the number of minutes he will have run by Day 5.

Running Schedule					
Day	1	2	3	4	5
Time (min)	30				

Independent Practice

Identify, describe, and extend each pattern.

4. 3, 5, 7, 9, 11, _____

The pattern is _____.

5. 26, 30, 34, 38, 42, _____

The pattern is _____.

6. 8, 8, 6, 6, 4, _____

The pattern is _____

_____.

7. 28, 24, 28, 24, 28, _____

The pattern is _____

_____.

8. 10, 20, 30, 40, 50, _____

The pattern is _____.

9. 3, 6, 12, 15, 21, _____

The pattern is _____.

Find a rule and extend each pattern.

10. The table shows the number of puppets sold at a toy store. If the pattern continues, how many puppets will be sold on Day 5?

Puppets Sold					
Day	1	2	3	4	5
Puppets Sold	7	5	9	7	

The pattern is _____

_____.

So, _____ puppets will be sold on Day 5.

11. Ria uses blocks to build model towers. The table shows how many blocks she needs for different-sized towers. How many blocks will she need to build a 7-foot tower?

Towers	
Tower Height (ft)	**Blocks Needed**
4	128
5	160
6	192
7	

_____ blocks

Algebra Find the unknown in each pattern.

12. 24, 29, _____, 39

13. 63, _____, 47, 39

14. _____, 17, 21, 25

15. _____, 86, 82, 84, 80, 82, 78

Problem Solving

16. It is recommended that a person drink 64 fluid ounces of water each day. The pattern shows how many days it would take to drink 448 fluid ounces. Explain how another rule could be used to find the same answer.

Recommended Water in a Week							
Day	1	2	3	4	5	6	7
Amount (fl oz)	64	128	192	256	320	384	448

+64 +64 +64 +64 +64 +64

My Work!

Mathematical
17. PRACTICE 8 Look for a Pattern The table at the right shows how much money Russell has each day.

Describe the rule for the pattern shown.

Overall, does Russell spend or earn more money?

Russell's Budget	
Day	**Russell's Money**
Monday	$35
Tuesday	$31
Wednesday	$27
Thursday	$32
Friday	$28
Saturday	$24
Sunday	$29

HOT Problems

Mathematical
18. PRACTICE 6 Explain to a Friend Create a number pattern involving two operations. Explain your pattern to a friend.

19. Building on the Essential Question Why is it important to look at more than just the first two numbers of a pattern to decide the rule for a pattern?

MY Homework

Homework Helper

Need help? ⟋ connectED.mcgraw-hill.com

Identify, describe, and extend the pattern.

12, 7, 14, 9, 16, 11

 Find the pattern by looking at how the numbers change. To get from 12 to 7, subtract 5. Then, to get from 7 to 14, add 7. See if the rule holds true for the rest of the numbers.

$$-5 \quad +7 \quad -5 \quad +7 \quad -5$$
$$12, \quad 7, \quad 14, \quad 9, \quad 16, \quad 11$$

2 The rule "subtract 5, add 7" does hold true. Now extend the pattern.

$$+7 \quad -5 \quad +7 \quad -5$$
$$12, 7, 14, 9, 16, 11, \quad 18, \quad 13, \quad 20, \quad 15$$

Practice

Identify, describe, and extend each pattern.

1. 39, 40, 36, 37, 33, 34, _____

The pattern is _____.

2. 64, 55, 46, 37, 28, 19, _____

The pattern is _____.

3. 53, 49, 52, 48, 51, 47, _____

The pattern is _____.

Problem Solving

Find a rule and extend each pattern.

4. **Mathematical PRACTICE 2 Use Number Sense** Duane and Mick are saving their money to build a tree house. Duane adds $5 to their piggy bank every other week. Mick adds $2 every week. So far they have saved $45. How many weeks have they been saving their money?

_____

5. Janelle practices basketball every afternoon in her driveway. Each day, her goal is to make 4 more baskets than she made the day before. If she makes 10 baskets the first day and meets her goal for 2 weeks, how many baskets will Janelle make on the 14th day?

_____

6. For every 5 coins Carmen gets, she gives 2 to her brother Frankie. If Carmen has 9 coins, how many does Frankie have?

_____

7. Kelli calls her grandmother every month. Every other month, Kelli also calls her cousin. If Kelli calls her cousin in January, how many calls will Kelli have made to her grandmother and her cousin by the end of August?

_____

Vocabulary Check

Draw a line to match the word to its example.

8. rule • 3, 7, 5, 9, 7, 11, 9

9. numeric pattern • add 4, subtract 2

Test Practice

10. Which pattern follows the rule "subtract 3, add 6"?

Ⓐ 18, 15, 21, 18, 24   Ⓒ 18, 15, 21, 18, 15

Ⓑ 18, 21, 15, 18, 12   Ⓓ 18, 24, 21, 27, 24

Sequences

Patterns follow a rule. Each number in a numeric pattern is called a **term**. The ordered arrangement of terms that make up a pattern is called a **sequence**.

 ## Math in My World Tutor

Example 1

Crystal starts reading her book on Monday. She reads 25 pages on the first day. Each day, she reads 25 pages. How many total pages will she have read by Tuesday, Wednesday, Thursday, and Friday?

The first term of the sequence is 25.

The rule is add 25.

Extend the pattern.

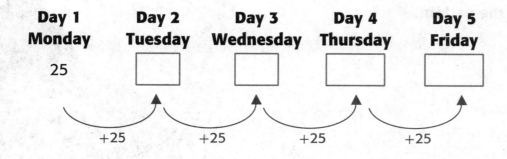

Day 1 Monday	Day 2 Tuesday	Day 3 Wednesday	Day 4 Thursday	Day 5 Friday
25	□	□	□	□

+25   +25   +25   +25

So, Crystal will have read _____ pages by Tuesday, _____ pages by Wednesday, _____ pages by Thursday, and _____ pages by Friday.

Example 2

The first term of a sequence is 65. The rule of the sequence is subtract 4. Find the next four terms in the sequence. Then make observations about the pattern.

 Find the next four terms.

Terms:	Term 1	Term 2	Term 3	Term 4	Term 5
Sequence:	65	☐	☐	☐	☐

$-4 \quad -4 \quad -4 \quad -4$

The next four terms in the sequence are ____, ____, ____, and ____.

② **Make observations about the pattern.**

Circle whether the terms are all odd or even. odd   even

Circle whether the terms increase or decrease. increase   decrease

Extend the pattern to a total of 10 terms.

65, ____, ____, ____, ____, ____, ____, ____, ____,

Make another observation about the pattern.

The ones digits repeat the pattern 5, 1, ____, ____, and ____.

Guided Practice

Extend each pattern by four terms. Write an observation about the pattern.

1. Rule: add 7

Pattern: 8, ____, ____, ____, ____

Observation: _____

_____

2. Rule: subtract 10

Pattern: 90, ____, ____, ____, ____

Observation: _____

_____

Talk MATH

How does the operation of a rule affect the terms of a sequence?

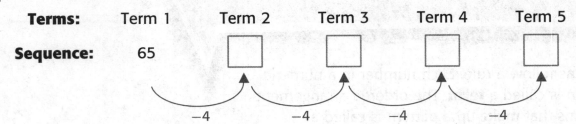

Independent Practice

Extend each pattern by four terms. Write an observation about the pattern.

3. Rule: add 9

Pattern: 7, ____, ____, ____, ____

Observation: _____

_____

4. Rule: add 12

Pattern: 2, ____, ____, ____, ____

Observation: _____

_____

5. Rule: subtract 9

Pattern: 87, ____, ____, ____, ____

Observation: _____

_____

6. Rule: subtract 5

Pattern: 86, ____, ____, ____, ____

Observation: _____

_____

7. Rule: multiply by 3

Pattern: 2, ____, ____, ____, ____

Observation: _____

_____

8. Rule: multiply by 4

Pattern: 5, ____, ____, ____, ____

Observation: _____

_____

9. Rule: divide by 2

Pattern: 64, ____, ____, ____, ____

Observation: _____

_____

10. Rule: divide by 5

Pattern: 625, ____, ____, ____, ____

Observation: _____

_____

11. Refer to the sequence 11, 16, 21, 26, 31, 36. Explain why the terms in the sequence will continue to alternate between even and odd numbers.

_____

_____

_____

Problem Solving

12. Each pumpkin costs $8. Jaime has already bought $24 worth of pumpkins. Suppose he buys five more pumpkins. How much will he spend in all after he buys each pumpkin? Write a sequence.

_____

13. 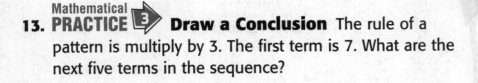 **Mathematical PRACTICE 3 Draw a Conclusion** The rule of a pattern is multiply by 3. The first term is 7. What are the next five terms in the sequence?

_____

Write two observations you can make about the pattern.

_____

_____

HOT Problems

14. **Mathematical PRACTICE 2 Use Number Sense** Write a sequence with at least 5 terms that forms a pattern. Identify the rule.

_____

_____

15. **Mathematical PRACTICE 2 Reason** The first term in a sequence is an odd number. The rule is to multiply by 2. Explain why the rest of the terms in the sequence will be even numbers.

_____

_____

16. **? Building on the Essential Question** How can I find patterns?

_____

_____

MY Homework

Lesson 3

Sequences

Homework Helper

Need help? connectED.mcgraw-hill.com

Extend the pattern described below by four terms. Then, note two observations about the pattern.

Use repeated subtraction to extend the pattern.

First Term: 46

$$
\begin{array}{cccc}
46 & 39 & 32 & 25 \\
\underline{-7} & \underline{-7} & \underline{-7} & \underline{-7} \\
39 & 32 & 25 & 18
\end{array}
$$

Rule: Subtract 7

So, the sequence is 46, 39, 32, 25, and 18.
The terms in the sequence decrease. The terms in the sequence also alternate between even and odd numbers.

Practice

Extend each pattern by four terms. Write an observation about the pattern.

1. Rule: add 8

Pattern: 5, _____, _____, _____, _____

Observation: _____

2. Rule: multiply by 2

Pattern: 3, _____, _____, _____, _____

Observation: _____

3. Rule: subtract 20

Pattern: 175, _____, _____, _____, _____

Observation: _____

4. Extend the pattern below by four terms. Write an observation about the pattern.

Rule: multiply by 10

Pattern: 26, _____, _____, _____, _____

Observation: _____

_____

Problem Solving

Mathematical
5. PRACTICE 8 **Look for a Pattern** Brad puts an equal amount of money in his savings account once a month. He started with $25. The next month, he had $35 in his account. Two months after that, he had $55 in his account. How much money will Brad have in his account after 6 months? Describe a rule. Then solve.

_____

6. On Monday, a toy store sold 4 race cars. On Tuesday, it sold 8 race cars. On Wednesday, it sold 16 race cars. Suppose this pattern continues. How many race cars will be sold on Friday? Describe a rule. Then solve.

_____

Vocabulary Check

Write a vocabulary word to complete each sentence.

sequence   term

7. Each number in a numeric pattern is a _____.

8. A _____ is the ordered arrangement of terms that make up a pattern.

Test Practice

9. Identify the next term in the sequence.   171, 141, 111, 81, _____

Ⓐ 61   Ⓑ 51   Ⓒ 41   Ⓓ 31

Operations and Algebraic Thinking
4.OA.5

CCSS

Problem-Solving Investigation

STRATEGY: Look for a Pattern

Lesson 4

ESSENTIAL QUESTION
How are patterns used in mathematics?

Learn the Strategy

 Watch Tutor

Daniel is training for a walk-a-thon. In the first week, he walked a total of five miles. In the second week, he walked a total of 7 miles. In the third week, he walked a total of 9 miles. Based on his pattern, how many miles will he walk in the fourth week?

1 Understand

What facts do you know?

Daniel walked _____ miles the first week, _____ miles the second week, and _____ miles the third week.

What do you need to find?

the number of miles Daniel will walk the _____ week

2 Plan

I will look for a pattern to solve the problem.

3 Solve

The sequence of the pattern is: 5, 7, 9.

The rule to the pattern is _____ .

Based on the rule, the next term in the sequence is _____ .

So, Daniel will walk _____ miles during the fourth week.

4 Check

Does your answer make sense? Explain.

....

Practice the Strategy

Taryn made 15 hair ribbons on Monday, 21 hair ribbons on Tuesday, and 27 hair ribbons on Wednesday. Based on her pattern, how many hair ribbons will she make on Thursday?

① Understand

What facts do you know?

_____

_____

What do you need to find?

_____

② Plan

_____

③ Solve

④ Check

Does your answer make sense? Explain.

_____

_____

Apply the Strategy

Solve each problem by looking for a pattern.

1. **Mathematical PRACTICE 8 Look for a Pattern** A store sold 48 model airplanes in August, 58 model airplanes in September, and 68 model airplanes in October. Suppose this pattern continues. How many model airplanes will be sold in December?

My Work!

2. The table shows how many tickets were sold for the school play each day.

Day	Number of Tickets
Monday	312
Tuesday	316
Wednesday	320
Thursday	324

Based on the pattern, how many tickets will be sold on Friday?

3. There are 80 picnic tables at the park. During the first weekend of the summer, there were 40 available tables. During the second weekend, there were 20 available tables. During the third weekend, there were 10 available tables. Based on the pattern, how many picnic tables will be available during the fourth weekend of the summer?

Review the Strategies

Use any strategy to solve each problem.

- Check for reasonableness.
- Make a table.
- Make a model.
- Look for a pattern.

4. Mathematical **PRACTICE 2** **Use Symbols** A theater can hold 200 people. Two groups rented out the theater. The first group has 92 people and the other has 107 people. Are there enough seats for everyone? Use symbols to explain.

My Work!

5. In one hour, Frank earns the money shown below. How much does he earn in 7 weeks if he works 3 hours each week?

6. What is the next number in the pattern 2, 5, 11, 23, ▇?

7. Katie sold 153 purses at a craft fair. How much money did she earn if each purse cost the amount shown below?

MY Homework

Homework Helper

Need help? ☞ connectED.mcgraw-hill.com

Solve the problem by looking for a pattern.

At ShopSmart, customers receive a $3 coupon if they spend $20, a $6 coupon if they spend $40, and a $9 coupon if they spend $60. If a customer spends $80, how many dollars worth of coupons will he or she receive if the pattern continues?

1 Understand

What facts do you know?

Customers get a $3 coupon for every $20 they spend.

What do you need to find?

I need to find how many dollars in coupons a customer gets for spending $80.

2 Plan

I will find a pattern to solve the problem.

3 Solve

The sequence of the pattern of the coupons is $3, $6, and $9.

The rule for the pattern is +$3.

Based on the rule, the next term in the sequence is $12.

So, customers receive a $12 coupon for spending $80.

4 Check

Does the answer make sense?
$9 + $3 = $12, so the answer makes sense.

Problem Solving

Solve each problem by looking for a pattern.

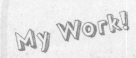

1. **Mathematical PRACTICE 1** **Plan Your Solution** Angela opened a new bakery. She got orders for 2 cakes the first week, 4 cakes the second week, and 8 cakes the third week. If the pattern continues, how many cake orders will Angela get the fourth week?

2. Manuel saw the following birds this week: 2 blue jays on Monday, 5 cardinals on Tuesday, 4 blue jays on Wednesday, 7 cardinals on Thursday, and 6 blue jays on Friday. If the pattern continues, what is the number and type of bird Manuel will see Saturday?

3. The house numbers on the north side of Flynn Street are even. In one block, the house numbers begin 1022, 1032, 1042, and 1052. What is the number of the next house likely to be?

4. Jessica is voting for dancers in a contest. She votes for the first contestant 5 times, the second contestant 9 times, and the third contestant 13 times. If she continues this pattern, how many times will Jessica vote for the fourth contestant?

5. A towel has a repeating pattern of 2 green stripes, then 3 blue stripes, and then 1 yellow stripe. If the towel has 20 stripes altogether, how many blue stripes are on it?

Vocabulary Check

Identify each pattern as a nonnumeric pattern or a numeric pattern.

1.

_____

2. 43, 46, 47, 50, 51, 54, 55

_____

3. 98, 88, 78, 68, 58, 48

_____

4.

_____

Use the pattern below for Exercises 5–7.

2, 6, 18, 54, 162

5. Put a circle around one **term** in the pattern.

6. Underline the **sequence**.

7. Write the **rule** for this pattern. _____

Concept Check

8. Extend the pattern. Draw the shapes on the lines.

 _____ _____

9. Identify, describe, and extend each pattern.

3, 8, 13, 18, 23, _____   The pattern is _____ .

10. Extend the pattern below by four terms. Write an observation about the pattern.

Rule: subtract 6

Pattern: 76, _____ , _____ , _____ , _____

Observation:

_____

_____

 ## Problem Solving

11. On Mondays, Wednesdays, and Fridays, Luke packs his lunch. On Tuesdays and Thursdays, Luke buys his lunch. Draw a nonnumeric pattern to show the pattern of Luke's lunch for two weeks.

12. Bob swims 10 laps on even numbered dates. He swims 15 laps on odd numbered dates. How many laps has he completed by the sixth of the month?

Test Practice

13. A nonnumeric pattern is shown below.

Which shows the next three objects in the pattern?

Ⓐ

Ⓑ

Ⓒ

Ⓓ

Addition and Subtraction Rules

Lesson 5

ESSENTIAL QUESTION
How are patterns used
in mathematics?

You can use a rule to write an equation that describes a pattern between **input** and **output** numbers. Tables can be used to show how input numbers change in the same way each time, creating a new output number.

 ## Math in My World

Example 1

Mr. Mathis is creating a table to show how input numbers are changed. Write an equation that describes the pattern in the table. Complete the table.

Pattern: $2 + \underline{} = 9$

$4 + \underline{} = 11$

$6 + \underline{} = 13$

Input (x)	Output (y)
2	9
4	11
6	13
8	
10	
12	

Rule: Add $\underline{}$.

Equation: $x + \underline{} = y$

⬆ Input   ⬆ Output

Use the rule to complete the table.

So, the equation that describes the pattern is $\underline{}$.

Example 2

A pizza shop offers $3 off any order over $10. Use the rule and equation to find the next four output numbers.

Input (c)	Output (d)
$11	$8
$12	
$14	
$16	
$18	

Rule: Subtract 3.

Equation: $c - \$3 = d$

Input → Output

Find the next four numbers when the input c is $12, $14, $16, and $18.

$c - \$3 = d$	$c - \$3 = d$	$c - \$3 = d$	$c - \$3 = d$
$12 - \$3 = \$\square$	$14 - \$3 = \$\square$	$16 - \$3 = \$\square$	$18 - \$3 = \$\square$

So, the next four amounts are _____.

Describe another pattern you see in this chart.

_____

_____

Guided Practice

1. Write an equation that describes the pattern. Then use the equation to find the next three output numbers.

Input (a)	5	9	13	17	21	25
Output (b)	9	13	17			

_____

Talk MATH

Explain what you should do if you test a number in an equation and it does not work.

Independent Practice

Write an equation that describes the pattern. Then use the equation to find the next two output numbers.

2.

Input (m)	11	16	21	26	31
Output (n)	2	7	12		

3.

Input (s)	2	6	10	14	18
Output (t)	15	19	23		

Equation: _____

Equation: _____

Use the rule to find the next four output numbers.

4.

Rule: $f + 3 = h$	
Input (f)	Output (h)
3	6
6	
9	
12	
15	

5.

Rule: $v - 11 = w$	
Input (v)	Output (w)
16	5
22	
28	
34	
40	

6.

Rule: $g - 5 = h$	
Input (g)	Output (h)
14	9
19	
24	
29	
34	

Create an input/output table for each equation.

7. $y + 4 = z$

Rule: $y + 4 = z$

8. $a - 7 = c$

Rule: $a - 7 = c$

9. Describe a pattern you see in Exercise 2.

_____

Problem Solving

The table shows what a taxi company charges
in dollars *c* for every *m* miles traveled.

10. Mathematical **PRACTICE** ② **Use Algebra** Use the table to write
an equation for this situation.

_____

11. Find the costs of a 25-mile trip and a 30-mile trip.

_____

12. Use the equation you wrote for Exercise 10 to find
the cost of a 60-mile trip.

_____

13. A different taxi company uses the equation
$c = m + \$4$ to determine their charges. Find the
cost of a 15-mile trip.

_____

Taxi Rates	
Input (*m*)	Output (*c*)
10	$12
15	$17
20	$22
25	⬛
30	⬛

Where to, buddy?

HOT Problems

14. Mathematical **PRACTICE** ④ **Model Math** Write a real-world situation that
can be represented by the table.

Input (*h*)	1	2	3	4	5
Output (*m*)	$10	$20	$30	⬛	⬛

_____

_____

15. ❓ **Building on the Essential Question** How can I find
the rule of a pattern?

_____

_____

MY Homework

Homework Helper

Need help? connectED.mcgraw-hill.com

**Write an equation that describes the pattern in the table.
Then use the equation to find the next three output numbers.**

Input (d)	12	15	18	21	24	27
Output (f)	19	22	25	■	■	■

The rule is add 7. The letter d represents the input, and the
letter f represents the output. So, the equation is $d + 7 = f$.

Use the equation to find the next three output numbers:

$21 + 7 = 28$
$24 + 7 = 31$
$27 + 7 = 34$

So, the completed table looks like this:

Input (d)	12	15	18	21	24	27
Output (f)	19	22	25	28	31	34

Practice

**Write an equation that describes the pattern. Then use the
equation to find the next three output numbers.**

1.

Input (a)	Output (b)
$2	$27
$4	$29
$6	
$8	
$10	

2.

Input (s)	Output (t)
87	76
80	69
73	
66	
59	

Equation: _____

Equation: _____

Write an equation that describes the pattern. Then use the equation to find the next three output numbers.

3.

Input (x)	Output (y)
22	17
26	21
30	
34	
38	

Equation: _____

4.

Input (c)	Output (d)
0	8
5	13
10	
15	
20	

Equation: _____

Problem Solving

Jeremy's class is going on a field trip. The school will bring all the students who are there that day plus 4 chaperones.

5. Write an equation for this situation.

Input (s)	Output (p)

6. Mathematical PRACTICE 5 Use Math Tools Complete the table to show how many people will go if there are 25, 27, 29, 31, or 33 students.

Vocabulary Check [Vocab]

Draw a line to match each word to its meaning.

7. input   • a number before an operation is performed

8. output   • a number that is the result of an operation

Test Practice

9. Refer to the equation $a - 6 = b$. If $a = 45$, what is the value of b?

Ⓐ 16   Ⓒ 51

Ⓑ 39   Ⓓ 60

Operations and Algebraic Thinking
4.OA.5

CCSS

Multiplication and Division Rules

You can write a multiplication or division equation
to extend a pattern.

Math in My World

Watch   Tutor

Example 1

**Charles washes cars to earn money. If he washes
2 cars, he earns $12. If he washes 6 cars, he earns
$36. Write an equation to describe the pattern.
Then use the equation to find how much money
Charles will earn if he washes 8, 10, and 12 cars.**

Complete the table. Then look for
the pattern that describes a rule.

Cars Washed	Amount Earned ($)
Input (*a*)	Output (*b*)
2	12
4	24
6	36
8	
10	
12	

Pattern: $2 \times \underline{\hspace{1cm}} = 12$

$4 \times \underline{\hspace{1cm}} = 24$

$6 \times \underline{\hspace{1cm}} = 36$

Rule: Multiply by \underline{\hspace{1cm}} .

Equation: $a \times \underline{\hspace{1cm}} = b$

 Input   Output

Find the next three output numbers when the input *a* is 8, 10, and 12.

$a \times 6 = b$ $a \times 6 = b$ $a \times 6 = b$

$8 \times 6 = \underline{\hspace{1cm}}$ $10 \times 6 = \underline{\hspace{1cm}}$ $12 \times 6 = \underline{\hspace{1cm}}$

So, Charles will earn $ \underline{\hspace{1cm}} , $ \underline{\hspace{1cm}} , and $ \underline{\hspace{1cm}} .

Example 2

Tutor

It costs $4 for each box of crackers. The equation is shown below. Use the equation to complete the table.

Total Cost ($)	Boxes of Crackers
Input (g)	Output (h)
4	1
8	
12	
16	
20	
24	

Rule: Divide by 4.

Equation: $g \div 4 = h$

           ↑   ↑

    (Input) (Output)

Find the next five output numbers when the input *g* is 8, 12, 16, 20, and 24.

$g \div 4 = h$	$g \div 4 = h$	$g \div 4 = h$	$g \div 4 = h$	$g \div 4 = h$
$8 \div 4 =$ _____	$12 \div 4 =$ _____	$16 \div 4 =$ _____	$20 \div 4 =$ _____	$24 \div 4 =$ _____

So, the next five output numbers are _____.

Describe another pattern you see in this table.

_____

_____

Talk MATH

How are a rule and an equation alike? How are they different?

Guided Practice

Check ✓

1. Write an equation that describes the pattern. Then use the equation to find the next three output numbers.

Input (w)	2	4	6	8	10	12
Output (v)	12	24	36			

Equation: _____

Describe a pattern you see in this chart.

_____

_____

Independent Practice

Write an equation that describes the pattern. Then use the equation to find the next three output numbers.

2.

Input (*m*)	1	3	5	7	9	11
Output (*n*)	5	15	25			

Equation: _____

3.

Input (*b*)	2	4	6	8	10	12
Output (*c*)	14	28	42			

Equation: _____

4.

Input (*j*)	4	8	12	16	20	24
Output (*k*)	1	2	3			

Equation: _____

5.

Input (*e*)	10	20	30	40	50	60
Output (*f*)	2	4	6			

Equation: _____

6.

Input (*x*)	16	24	32	40	48	56
Output (*y*)	2	3	4			

Equation: _____

7.

Input (*t*)	12	10	8	6	4	2
Output (*v*)	24	20	16			

Equation: _____

Create an input/output table for each equation.

8. $a \times 5 = b$ Rule: $a \times 5 = b$

9. $c \div 6 = d$ Rule: $c \div 6 = d$

10. Describe a pattern you see in Exercise 6.

_____

_____

Problem Solving

Sari makes bead necklaces. The table shows the number of blue beads and green beads Sari uses.

Blue Beads	Green Beads
Input (j)	Output (k)
3	1
9	3
15	5
21	▪
27	▪
33	▪

11. **Mathematical PRACTICE 2** **Use Algebra** Write an equation that describes the relationship between green beads and blue beads.

_____

12. How many green beads does Sari need if she is using 36 blue beads?

_____

13. How many beads does Sari have in all if she has 9 green beads?

_____

My Work!

HOT Problems

14. **Mathematical PRACTICE 2** **Reason** Circle the operation that can be used to write an equation for the input/output table to the right. Explain.

Input (m)	Output (n)
1	2
2	4
3	6

addition   subtraction   multiplication   division

_____

_____

15. **Building on the Essential Question** How can an input/output table help me solve a real-world problem?

_____

_____

MY Homework

Homework Helper

Need help? ⟋ connectED.mcgraw-hill.com

Write an equation that describes the pattern in the table below. Then use the equation to find the next three output numbers.

Input (k)	2	4	6	8	10	12
Output (m)	6	12	18	■	■	■

The rule is multiply by 3. The letter k represents the input, and the letter m represents the output. So, the equation is $k \times 3 = m$.

Use the equation to find the next three output numbers:

$8 \times 3 = 24$
$10 \times 3 = 30$
$12 \times 3 = 36$

So, the completed table looks like this:

Input (k)	2	4	6	8	10	12
Output (m)	6	12	18	24	30	36

Practice

Write an equation that describes the pattern. Then use the equation to find the next three output numbers.

1.

Input (a)	Output (b)
7	1
14	2
21	
28	
35	

Equation: _____

2.

Input (s)	Output (t)
99	33
84	28
69	
54	
39	

Equation: _____

Write an equation that describes the pattern. Then use the equation to find the next three output numbers.

3.

Input (x)	Output (y)
$5	$40
$6	$48
$7	
$8	
$9	

Equation: _____

4.

Input (c)	Output (d)
50	10
45	9
40	
35	
30	

Equation: _____

Problem Solving

Shawna found out there are 4 yellow pencils for every blue pencil.

5. **Mathematical PRACTICE 2 Use Algebra** Write an equation for this situation.

_____

Input (s)	Output (p)

6. Complete the table to show how many yellow pencils there are if there are 5, 7, 9, 11, or 13 blue pencils.

Test Practice

7. Refer to the equation $a \times 9 = b$. If $a = 3$, what is the value of b?

(A) 3   (B) 12   (C) 18   (D) 27

Order of Operations

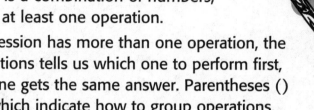

Copyright © The McGraw-Hill Companies, Inc.

Lesson 7

ESSENTIAL QUESTION
How are patterns used
in mathematics?

An expression is a combination of numbers, variables, and at least one operation.

When an expression has more than one operation, the order of operations tells us which one to perform first, so that everyone gets the same answer. Parentheses () are symbols which indicate how to group operations.

 ## Math in My World [Watch ▶] [Tutor 💬]

Example 1

The table shows how much movie tickets cost. How much will it cost to buy 3 adult tickets and 5 child tickets?

Ticket	Cost
Adult	$8
Child	$5

cost = 3 adult tickets + 5 child tickets

$c = 3 \times \$8 \ + \ 5 \times \$5$

$c = \$\boxed{} + \$\boxed{}$ First, multiply 3 by $8 and 5 by $5.

$c = \ \$\boxed{}$ Add the products to find the total cost.

So, the total cost is $ _____ .

Key Concept Order of Operations

1. Perform operations in parentheses.
2. Multiply and divide in order from left to right.
3. Add and subtract in order from left to right.

Remember, an expression is a combination of numbers, variables, and at least one operation.

Example 2

Find the value of the expression 3 × (4 + 6).

$3 \times (4 + 6)$

$3 \times \boxed{}$ Perform the operation in the parentheses first.

$\boxed{}$ Multiply.

So, $3 \times (4 + 6) = $ _____ .

Example 3

Find the value of the expression (7 − 3) ÷ (2 + 2).

$(7 - 3) \div (2 + 2)$

$\boxed{} \div \boxed{}$ Perform the operations in the parentheses first.

$\boxed{}$ Divide.

So, $(7 - 3) \div (2 + 2) = $ _____ .

Talk MATH

Explain why Exercises 2 and 3 have different answers even though the numbers are the same.

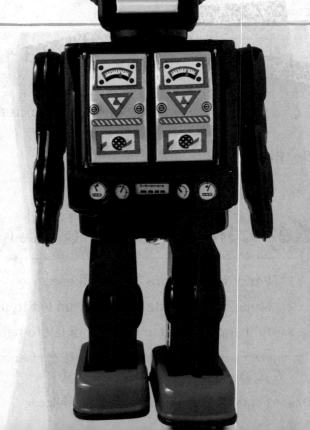

Guided Practice

Find the value of each expression.

1. $12 - 1 \times 3 = $ _____

2. $15 - 4 \times 2 = $ _____

3. $(15 - 4) \times 2 = $ _____

Independent Practice

Find the value of each expression.

4. $8 + 5 \times 2 =$ _____

5. $10 - 1 \times 5 =$ _____

6. $4 + 6 \div 2 =$ _____

7. $9 \times 2 - 6 =$ _____

8. $(16 + 2) \div 3 =$ _____

9. $6 \times (6 - 2) =$ _____

10. $(12 - 4) \div 4 =$ _____

11. $12 - (4 \div 4) =$ _____

12. $(3 + 6) \div (3 \times 1) =$ _____

13. $3 + (6 \div 3) \times 1 =$ _____

Algebra **Use the order of operations to find the unknown in each equation.**

14. $5 \times 4 - \blacksquare = 13$

The unknown is _____.

15. $\blacksquare \times (8 + 6) = 42$

The unknown is _____.

16. $(2 + 1) \times (9 - \blacksquare) = 12$

The unknown is _____.

17. $(10 \div 2) + (\blacksquare + 3) = 40$

The unknown is _____.

Problem Solving

18. Each bag of dried apples has 5 servings. Each bag of dried apricots has 3 servings. How many servings of dried fruit are in 6 bags of dried apples and 2 bags of dried apricots?

_____

19. Each book costs $4. How much does it cost to buy 3 books and one magazine that costs $5?

_____

20. **Mathematical PRACTICE** **2** **Use Number Sense** A sandwich costs $6 and a drink costs $3. How much does it cost to buy 4 sandwiches and 4 drinks?

_____

My Work!

HOT Problems

21. **Mathematical PRACTICE** **1** **Keep Trying** Use each of the numbers 1, 2, 3, and 4 exactly once in the equation below to make the equation true.

$$(\boxed{} \times \boxed{}) + (\boxed{} \div \boxed{}) = 10$$

22. **Mathematical PRACTICE** **1** **Make a Plan** Find possible unknown values to make the equation true.

$$(\boxed{} \times \boxed{}) + (\boxed{} \div 2) = 15$$

23. **? Building on the Essential Question** Why is knowing the order of operations important?

_____

_____

_____

MY Homework

Homework Helper eHelp

Need help? connectED.mcgraw-hill.com

Find the value of each expression.

8 × 7 − (9 ÷ 3) = ?

8 × 7 − (9 ÷ 3)

8 × 7 −   3   Perform the operations in parentheses first.

56   −   3   Multiply.

53   Subtract.

So, 8 × 7 − (9 ÷ 3) = 53.

---

24 − 2 + 6 × 3 = ?

24 − 2 + 6 × 3

24 − 2 +   18   Multiply.

22   +   18   Subtract.

40   Add.

So, 24 − 2 + 6 × 3 = 40.

Practice

Find the value of each expression.

1. 5 + 9 ÷ 3 = _____

2. 46 − (6 × 5) = _____

Find the value of each expression.

3. $(3 + 1) + 27 \div 9 =$ _____

4. $5 \times 5 - 8 =$ _____

5. $(4 + 20) \div 2 + 6 =$ _____

6. $2 \times 9 + 14 \div 2 =$ _____

 Problem Solving

7. **Mathematical PRACTICE** **4** **Model Math** Tami buys two books that cost $14 each. She pays an additional $2 in tax. How much did Tami pay altogether?

_____

8. Claudio had 34 toy cars. He lost two at the park. Then he divided the rest of the cars evenly among himself and 3 cousins. How many cars did each child get?

_____

9. Last week, Jean did two sit-ups on Monday and three sit-ups on Wednesday. This week, Jean did three times as many sit-ups as last week. How many sit-ups did Jean do this week?

_____

Test Practice

10. Which expression has a value of 20?

   Ⓐ $2 \times 5 + 5$ Ⓒ $3 \times 7 - 1$

   Ⓑ $3 \times (5 + 5)$ Ⓓ $40 \div 5 - 3$

Check My Progress

Vocabulary Check

Use the words in the word bank to complete each sentence.

equation	input	operation
output	unknown	

1. In the equation $4 + x = 7$, the variable x is a(n) _____.

2. In the table to the right, the letter m

   represents the _____. The letter n

   represents the _____.

$m + 5 = n$	
m	n
2	7
3	8

3. A(n) _____ is a sentence
   that contains an equals sign ($=$), showing that two
   expressions are equal.

4. Addition is an example of a(n) _____.

Concept Check

**Write an equation that describes the pattern. Then use
the equation to find the next three output numbers.**

5.

Input (a)	4	5	6	7	8	9
Output (b)	9	10	11			

Equation: _____

6.

Input (c)	6	8	10	12	14	16
Output (d)	12	16	20			

Equation: _____

Find the value of each expression.

7. $(7 + 5) \div 3 =$ _____

8. $11 - 2 \times 5 =$ _____

Problem Solving

9. The amount in dollars c a bus company charges to take s students on a field trip are shown at the right. Write an equation to describe the pattern. Then complete the table to show how much it would cost for 40 and 50 students to go on a field trip.

Students	Cost ($)
10	60
20	70
30	80
40	
50	

My Work!

10. A local sports team sells 6 tickets for $3, 8 tickets for $4, and 10 tickets for $5. Write a rule and equation to find the cost of 20 tickets.

11. Each peanut butter snack costs $2. Each chocolate snack costs $3. How much does it cost to buy 6 peanut butter snacks and 8 chocolate snacks? Write an equation.

Test Practice

12. What is the value of m in the equation to the right if $n = 6$?

Ⓐ 15

Ⓒ 54

Ⓑ 27

Ⓓ 81

$$9 \times n = m$$

Hands On
Equations with Two Operations

Lesson 8

ESSENTIAL QUESTION
How are patterns used
in mathematics?

Sometimes an equation has more than one operation.

Build It

Model the equation $(n \times 3) + 5 = y$.

1. **Set up an equation machine.**
   Use paper plates to represent the variables,
   rubber bands for parentheses, paper clips
   for the equals sign, and index cards to show
   the numbers and operations.

2. **Input counters to find y.**
   Suppose n equals 1. Place 1 counter on the
   plate labeled n. Move the counters through
   the machine, following the operations given.
   Use the order of operations.

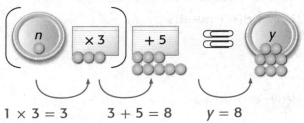

$1 \times 3 = 3$ $3 + 5 = 8$ $y = 8$

3. **Record the equation.**
   Fold a piece of paper in half and label it as shown.

when n is... y equals...
   1   8

So, when $n = 1$, $y =$ _____ .

Try It

Input more values for the equation $(n \times 3) + 5 = y$. Find new input and output values.

Suppose n equals 2. Find the value of y.

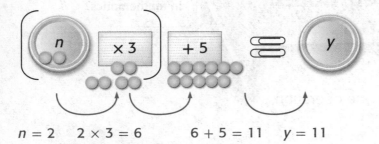

$n = 2$ $2 \times 3 = 6$ $6 + 5 = 11$ $y = 11$

So, when $n = 2$, $y =$ _____ .

Repeat the process for input values of 3, 4, and 5.

Record your input (n) and output (y) values.

When *n* is...	*y* equals...
1	8
2	
3	
4	
5	

Talk About It

1. **Mathematical PRACTICE 2** **Use Algebra** Refer to the equation $(n \times 3) + 5 = y$. What is the value of y when n equals 6? 7?

_____

2. Given the equation $(n + 7) \times 3 = y$, how would you find the value of y if n equals 3?

_____

_____

Practice It

Use each equation to find each unknown. Use models if needed.

3. $(t + 8) \times 2 = s$

When $t = 4$, $s =$ _____.

4. $(m \times 6) + 4 = d$

When $m = 3$, $d =$ _____.

5. $8 + (z \times 2) = w$

When $z = 5$, $w =$ _____.

6. $(a \div 6) + 5 = b$

When $a = 18$, $b =$ _____.

7. $12 - (e \times 4) = f$

When $e = 2$, $f =$ _____.

8. $(r + 8) \times 6 = s$

When $r = 3$, $s =$ _____.

9. $(g - 4) \times 7 = h$

When $g = 12$, $h =$ _____.

10. $64 \div (p + 4) = q$

When $p = 4$, $q =$ _____.

Apply It

11. Set up an equation machine to show $(x + 4) \div 3 = y$.
Find the values of y when $x = 8$, $x = 11$, and $x = 20$.

When $x = 8$, $y = $ _____.

When $x = 11$, $y = $ _____.

When $x = 20$, $y = $ _____.

12. **Mathematical PRACTICE 3** **Find the Error** Robert is finding
the output values for the equation $a + 7 \times 3 = b$.
He wrote a few statements about the input and
output values.

> When $a = 3$, $b = 30$.
>
> When $a = 4$, $b = 33$.
>
> When $a = 5$, $b = 36$.

What is Robert's mistake?

_____

_____

Use the equation correctly to find the output values.

> When $a = 3$, $b = $ _____.
>
> When $a = 4$, $b = $ _____.
>
> When $a = 5$, $b = $ _____.

Rewrite the equation so that Robert's values are correct.

_____

Write About It

13. How do parentheses affect the value of expressions?

_____

_____

_____

MY Homework

Homework Helper eHelp

Need help? connectED.mcgraw-hill.com

Use the equation $(t \times 3) + 5 = w$ to find w when $t = 4$.

$(t \times 3) + 5 = w$

$(4 \times 3) + 5 = w \qquad t = 4$

$12 + 5 = w \qquad$ Perform the operation inside parentheses.

$17 = w \qquad$ Add.

When $t = 4$, $w = 17$.

The counters at the right
model this equation.

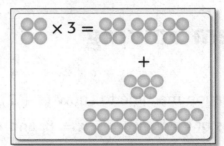

Practice

**Use each equation to find each unknown. Draw models
if needed.**

1. $(z + 3) \times 2 = y$

When $z = 2$, $y = $ _____.

2. $4 + (g \times 3) = m$

When $g = 3$, $m = $ _____.

Use each equation to find each unknown. Draw models if needed.

3. $2 + (n \times 7) = p$

When $n = 1$, $p =$ _____.

4. $(r \times 2) + 6 = v$

When $r = 4$, $v =$ _____.

5. $6 + (a \times 3) = b$

When $a = 5$, $b =$ _____.

6. $(j \div 4) + 8 = k$

When $j = 16$, $k =$ _____.

Problem Solving

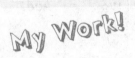

7. Set up an equation machine to show $(x + 2) \times 5 = y$. Find the values of y when $x = 5$, $x = 8$, and $x = 12$.

When $x = 5$, $y =$ _____.

When $x = 8$, $y =$ _____.

When $x = 12$, $y =$ _____.

Mathematical
8. PRACTICE 2 Understand Symbols Bryan made up a game. Each team starts with seven points. Each time a team answers a question correctly, they earn five points. The equation used to find the total number of points is $7 + (5 \times q) = t$. Find the total number of points (t) when a team answers six questions (q) correctly.

_____

Name _____

Equations with Multiple Operations

Lesson 9

ESSENTIAL QUESTION
How are patterns used in mathematics?

You have used tables to show equations with one operation. A table can also help you show equations with two operations.

 ## Math in My World Watch Tutor

Example 1

Sam earns $7 each time he rakes his neighbor's lawn. He also earns $5 each week for doing chores at home. Sam wants to find what he will earn in a week if he does chores and rakes the lawn 1, 2, or 3 times.

1 **Write an equation.**

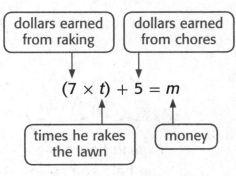

dollars earned from raking   dollars earned from chores

$$(7 \times t) + 5 = m$$

times he rakes the lawn   money

2 **Complete the table.**

So, during one week, if Sam rakes the lawn

once, he will earn $ _____ . If he rakes

the lawn twice, he will earn $ _____ .

If he rakes the lawn three times,

he will earn $ _____ .

Sam's Money		
Input (t)	(7 × t) + 5 = m	Output (m)
1	(7 × 1) + 5 = 12	12
2	(7 × 2) + 5 = 19	
3	(7 × 3) + 5 =	

Online Content at **connectED.mcgraw-hill.com**

Equations can have multiple operations.

Example 2

Complete the table to find output numbers when x = 2, 3, 4, and 5.

Helpful Hint
Solve what is between the parentheses first.

$2 \times (9 - x) + 3 = y$	
Input (x)	Output (y)
1	19
2	
3	
4	
5	

Find the value of y when x = 2.

$2 \times (9 - 2) + 3 = y$

$2 \times \quad 7 \quad + 3 = y$

$14 + 3 = y$

$17 \quad = y$

Repeat the process when x = 3, 4, and 5.

When x = 3, y = _____.

When x = 4, y = _____.

When x = 5, y = _____.

Describe patterns you see in the table.

_____

_____

_____

Talk MATH

Explain how tables can help you solve a problem.

Guided Practice Check ✓

1. Complete the table.

$(5 + x) \times 4 = y$	
Input (x)	Output (y)
1	24
2	
3	
4	

Independent Practice

Complete each table.

2.

$(7 - x) \times 7 = y$	
Input (x)	Output (y)
1	42
2	
3	
4	

3.

$(2 + x) \times 6 = y$	
Input (x)	Output (y)
1	18
2	
3	
4	

4.

$(4 \times x) - 3 = y$	
Input (x)	Output (y)
1	1
2	
3	
4	

5.

$(9 - x) + 2 = y$	
Input (x)	Output (y)
1	10
2	
3	
4	

6.

$(12 \div x) + 5 = y$	
Input (x)	Output (y)
1	17
2	
3	
4	

7.

$(14 - x) \div 2 = y$	
Input (x)	Output (y)
2	6
4	
6	
8	

8.

$(5 \times x) \div 5 + 1 = y$	
Input (x)	Output (y)
1	2
2	
3	
4	

9.

$3 \times (10 - x) + 4 = y$	
Input (x)	Output (y)
1	31
3	
5	
7	

Problem Solving

10. **Mathematical PRACTICE 5** **Use Math Tools** It costs $3 to park at the fair. Tickets cost $6 each. How much will it cost a family of 4 to go to the fair? Complete the table to solve.

($6 × x) + $3 = y	
Input (x)	Output (y)
1	$9
2	
3	
4	

_____

11. Tam walks 2 miles to school each day. During gym class, she always runs three times as far as Dante. How many miles will Tam walk and run if Dante runs 1 mile?

_____

HOT Problems

12. **Mathematical PRACTICE 3** **Find the Error** Ashley completed the table shown. Find and correct her mistake.

(10 − x) × 2 = y	
Input (x)	Output (y)
1	11
2	10
3	9

_____

_____

_____

_____

13. **Building on the Essential Question** Describe a real-world situation that could use a table with two operations.

_____

_____

MY Homework

Homework Helper

Need help? ⌐ connectED.mcgraw-hill.com

Lauren's recipe calls for 2 times as many cups of flour as sugar. She always adds 1 cup of oatmeal. If she uses 2, 3, or 4 cups of sugar, how many cups of flour and oatmeal will she use?

1 Write an equation.

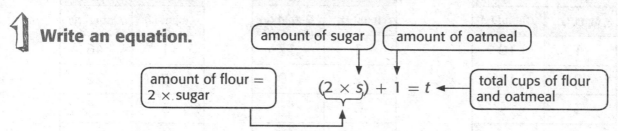

| amount of sugar | amount of oatmeal |

amount of flour = 2 × sugar

$(2 \times s) + 1 = t$ ← total cups of flour and oatmeal

2 Make a table.

$(2 \times s) + 1 = t$	
Input	**Output**
2	5
3	7
4	9

$(2 \times 2) + 1 = 5$
$(2 \times 3) + 1 = 7$
$(2 \times 4) + 1 = 9$

If she uses 2 cups of sugar, she will use 5 cups of flour and oatmeal.
If she uses 3 cups of sugar, she will use 7 cups of flour and oatmeal.
If she uses 4 cups of sugar, she will use 9 cups of flour and oatmeal.

Practice

1. Complete the table.

$(3 \times x) + 2 = y$	
Input (x)	**Output (y)**
1	5
2	8
3	
4	

Complete each table.

2.

$(12 \div x) + 3 = y$	
Input (x)	Output (y)
1	15
2	9
3	
4	

3.

$(4 + x) \times 6 = y$	
Input (x)	Output (y)
1	30
2	36
3	
4	

4.

$(10 - x) \times 7 = y$	
Input (x)	Output (y)
1	63
2	56
3	
4	

5.

$(5 \times x) + 5 = y$	
Input (x)	Output (y)
1	10
2	15
3	
4	

6.

$(6 + x) \times 2 + 3 = y$	
Input (x)	Output (y)
1	17
2	19
3	
4	

7.

$2 \times (24 \div x) - 2 = y$	
Input (x)	Output (y)
1	46
2	22
3	
4	

 # Problem Solving

8. Mathematical PRACTICE 1 Make Sense of Problems Mauricio hits a baseball 4 times as often as Tony each game. He also hits 20 baseballs every Monday at practice. How many baseballs will Mauricio hit this week if Tony hits 4 balls at the game Saturday?

9. Callie loves flowers. She picks 4 tulips for every daisy she picks. Callie's mom also gave her 6 tulips this week from her garden. How many tulips will Callie have this week if she picks 3 daisies?

Test Practice

10. Refer to the equation $(x \times 3) - 2 = y$. If $x = 7$, what is the value of y?

Ⓐ $y = 27$ Ⓑ $y = 23$ Ⓒ $y = 21$ Ⓓ $y = 19$

Review

Vocabulary Check

Write the letter of the definition next to the correct word.

1. equation _____

2. input _____

3. nonnumeric pattern _____

4. numeric pattern _____

5. operation _____

6. output _____

7. pattern _____

8. rule _____

9. sequence _____

10. term _____

11. unknown _____

A. A statement that describes a relationship between numbers or objects.

B. A pattern that uses numbers.

C. A sentence that contains an equals sign (=), showing that two expressions are equal.

D. An amount that is not known.

E. A mathematical process such as addition, subtraction, multiplication, or division.

F. A collection of terms that show a pattern.

G. A sequence showing a relationship among terms that are not numbers.

H. Each number in a sequence.

I. The result of an input quantity being changed by a function.

J. A sequence of terms that follow a certain order.

K. A quantity that is changed by a function to produce an output.

Concept Check ✓

12. Extend the pattern. Draw the shape on the line.

      _____

Identify, describe, and extend each pattern.

13. 4, 20, 100, 500, _____

The pattern is _____

_____

14. 44, 22, 20, 10, 8, _____

The pattern is _____

_____

Extend each pattern by four terms. Write an observation about the pattern.

15. Rule: add 6

Pattern: 3, ____, ____, ____, ____

Observation:

_____

_____

16. Rule: multiply by 2

Pattern: 4, ____, ____, ____, ____

Observation:

_____

_____

Write an equation that describes the pattern. Then use the equation to find the next two output numbers.

17.

Input (j)	25	35	45	55	65
Output (k)	21	31	41		

Equation: _____

18.

Input (g)	1	2	3	4	5
Output (h)	3	6	9		

Equation: _____

Find the value of each expression.

19. $7 + 3 \times 6 =$ _____

20. $(6 - 4) \times 9 =$ _____

Find the unknown.

21. $(f + 5) \times 3 = g$

When $f = 4$, $g =$ _____ .

22. $(x \times 4) + 7 = y$

When $x = 8$, $y =$ _____ .

Name _____

Problem Solving

23. Kenneth displays his picture frames in the pattern shown below. Every other picture is of his friends and the rest are of his family. If the first picture is of his family, what picture will be in the third square frame?

24. The admission for an art museum costs $5 per person. Complete the table to find how much it would cost for 2, 3, 4, 5, and 6 people to visit the museum.

People	Cost
2	
3	
4	
5	
6	

25. Mrs. Brown's class can earn 5 minutes of extra recess for each marble she puts in a jar. The class has 15 minutes of recess each morning. Make an input/output table to find how many minutes of recess they will get if they earn 3 marbles.

Test Practice

26. Find the value of the expression $(5 + 2) \times 7$.

- Ⓐ 14
- Ⓑ 19
- Ⓒ 21
- Ⓓ 49

Reflect

Use what you learned about patterns to
complete the graphic organizer.

ESSENTIAL QUESTION

How are patterns
used in mathematics?

Vocabulary	Nonnumeric Patterns	Numeric Patterns

Reflect on the ESSENTIAL QUESTION **Write your answer below.**

_____

_____

_____

Glossary/Glosario

← Go online for the eGlossary.

Go to the *eGlossary* to find out more about these words in the following 13 languages:

Arabic • Bengali • Brazilian Portuguese • Cantonese • English • Haitian Creole
Hmong • Korean • Russian • Spanish • Tagalog • Urdu • Vietnamese

Aa	**English**	**Spanish/Español**

acute angle   An *angle* with a measure greater than 0° and less than 90°.

ángulo agudo   *Ángulo* que mide más de 0° y menos de 90°.

acute triangle   A *triangle* with all three *angles* less than 90°.

triángulo acutángulo   *Triángulo* cuyos tres *ángulos* miden menos de 90°.

add (adding, addition)   An *operation* on two or more *addends* that results in a *sum*.

$$9 + 3 = 12$$

suma (sumar)   *Operación* de dos o más *sumandos* que da como resultado una *suma*.

$$9 + 3 = 12$$

addend   Any numbers being *added* together.

sumando   Cualquier número que se *suma* a otro.

algebra   A branch of mathematics that uses symbols, usually letters, to explore relationships between quantities.

álgebra   Rama de las matemáticas en la que se usan símbolos, generalmente letras, para explorar relaciones entre cantidades.

Copyright © The McGraw-Hill Companies, Inc.

Online Content at **connectED.mcgraw-hill.com**   Glossary/Glosario **GL1**

Aa

angle A figure that is formed by two *rays* with the same *endpoint*.

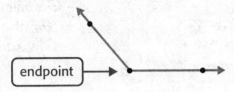

area The number of *square units* needed to cover the inside of a region or plane figure without any overlap.

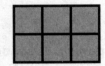

area = 6 square units

Associative Property of Addition The property that states that the grouping of the *addends* does not change the *sum*.

$$(4 + 5) + 2 = 4 + (5 + 2)$$

Associative Property of Multiplication The property that states that the grouping of the *factors* does not change the *product*.

$$3 \times (6 \times 2) = (3 \times 6) \times 2$$

ángulo Figura formada por dos *semirrectas* con el mismo *extremo*.

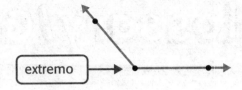

área Cantidad de *unidades cuadradas* necesarias para cubrir el interior de una región o figura plana sin superposiciones.

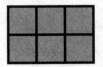

área = 6 unidades cuadradas

propiedad asociativa de la suma Propiedad que establece que la agrupación de los *sumandos* no altera la *suma*.

$$(4 + 5) + 2 = 4 + (5 + 2)$$

propiedad asociativa de la multiplicación Propiedad que establece que la agrupación de los *factores* no altera el *producto*.

$$3 \times (6 \times 2) = (3 \times 6) \times 2$$

Bb

bar graph A graph that compares *data* by using bars of different *lengths* or heights to show the values.

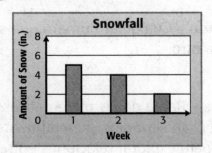

gráfica de barras Gráfica en la que se comparan los *datos* usando barras de distintas *longitudes* o alturas para mostrar los valores.

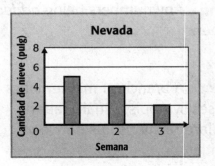

capacity The amount of liquid a container can hold.

centimeter (cm) A *metric* unit for measuring *length*.

100 centimeters = 1 meter

circle A closed figure in which all points are the same distance from a fixed point, called the center.

Commutative Property of Addition The property that states that the order in which two numbers are *added* does not change the *sum.*

$$12 + 15 = 15 + 12$$

Commutative Property of Multiplication The property that states that the order in which two numbers are *multiplied* does not change the *product.*

$$7 \times 2 = 2 \times 7$$

compatible numbers Numbers in a problem or related numbers that are easy to work with mentally.

720 and 90 are compatible numbers for *division* because $72 \div 9 = 8$.

composite number A whole number that has more than two factors.

12 has the factors 1, 2, 3, 4, 6, and 12.

capacidad Cantidad que puede contener un recipiente, medida en unidades de volumen.

centímetro (cm) Unidad métrica de *longitud.*

100 centímetros = 1 metro

círculo Figura cerrada en la cual todos los puntos equidistan de un punto fijo llamado centro.

propiedad conmutativa de la suma Propiedad que establece que el orden en el que se *suman* dos o más números no altera la *suma.*

$$12 + 15 = 15 + 12$$

propiedad conmutativa de la multiplicación Propiedad que establece que el orden en el que se *multiplican* dos o más números no altera el *producto.*

$$7 \times 2 = 2 \times 7$$

números compatibles Números en un problema o números relacionados con los cuales es fácil trabajar mentalmente.

720 ÷ 90 es una división que usa números compatibles porque $72 \div 9 = 8$.

número compuesto Número natural con más de dos factores.

12 tiene los factores 1, 2, 3, 4, 6 y 12.

Cc

congruent figures Two figures having the same size and the same shape.

figuras congruentes Dos figuras con la misma forma y el mismo tamaño.

convert To change one unit to another.

convertir Cambiar de una unidad a otra.

cup (c) A *customary* unit of *capacity* equal to 8 fluid ounces.

taza (tz) Unidad *usual* de *capacidad* que equivale a 8 onzas líquidas.

customary system The measurement system most often used in the United States. Units include *foot*, *pound*, and *quart*.

sistema usual Conjunto de unidades de medida de uso más frecuente en Estados Unidos. Incluye unidades como *el pie, la libra* y *el cuarto.*

Dd

data Numbers or symbols, sometimes collected from a *survey* or experiment, to show information. Datum is singular; data is plural.

datos Números o símbolos que muestran información, algunas veces recopilados a partir de una *encuesta* o un experimento.

decimal A number with one or more digits to the right of the decimal point, such as 8.37 or 0.05.

decimal Número con uno o más dígitos a la derecha del *punto decimal,* como 8.37 o 0.05.

decimal equivalents Decimals that represent the same number.

0.3 and 0.30

decimales equivalentes Decimales que representan el mismo número.

0.3 y 0.30

decimal point A period separating the ones and the *tenths* in a decimal number.

0.8 or $3.77

punto decimal Punto que separa las unidades de las *décimas* en un número decimal.

0.8 o $3.77

decompose To break a number into different parts.

descomponer Separar un número en diferentes partes.

degree (°) **a.** A unit for measuring angles. **b.** A unit of measure used to describe temperature.

grado (°) **a.** Unidad que se usa para medir ángulos. **b.** Unidad de medida que se usa para describir la temperatura.

denominator The bottom number in a *fraction*.

In $\frac{5}{6}$, 6 is the denominator.

digit A symbol used to write numbers. The ten digits are 0, 1, 2, 3, 4, 5, 6, 7, 8, and 9.

Distributive Property To *multiply* a *sum* by a number, multiply each *addend* by the number and *add* the *products*.

$$4 \times (1 + 3) = (4 \times 1) + (4 \times 3)$$

dividend A number that is being *divided*.

$3\overline{)19}$ 19 is the dividend.

division (divide) An *operation* on two numbers in which the first number is split into the same number of equal groups as the second number.

divisor The number by which the *dividend* is being *divided*.

$3\overline{)19}$ 3 is the divisor.

denominador El número de abajo en una *fracción*.

En $\frac{5}{6}$, 6 es el denominador.

dígito Símbolo que se usa para escribir los números. Los diez dígitos son 0, 1, 2, 3, 4, 5, 6, 7, 8 y 9.

propiedad distributiva Para *multiplicar* una *suma* por un número, multiplica cada *sumando* por ese número y luego *suma* los *productos*.

$$4 \times (1 + 3) = (4 \times 1) + (4 \times 3)$$

dividendo Número que se *divide*.

$3\overline{)19}$ 19 es el dividendo.

división (dividir) *Operación* entre dos números en la que el primer número se separa en tantos grupos iguales como indica el segundo número.

divisor Número entre el cual se *divide* el *dividendo*.

$3\overline{)19}$ 3 es el divisor.

elapsed time The amount of time that has passed from beginning to end.

endpoint The point at either end of a *line segment* or the point at the beginning of a *ray*.

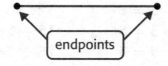

endpoints

tiempo transcurrido Cantidad de tiempo que ha pasado entre el principio y el fin de algo.

extremo Punto que se encuentra en cualquiera de los dos lados en que termina un *segmento de recta* o al principio de una *semirrecta*.

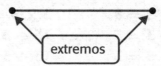

extremos

equation   A sentence that contains an equals sign (=), showing that two *expressions* are equal.

ecuación   Espresión matemática que contiene el signo igual, (=), que indica que las dos *expresiones* son iguales.

equilateral triangle   A *triangle* with three *congruent* sides.

triángulo equilátero   *Triángulo* con tres lados *congruentes*.

equivalent fractions   *Fractions* that represent the same number.

$$\frac{3}{4} = \frac{6}{8}$$

fracciones equivalentes   *Fracciones* que representan el mismo número.

$$\frac{3}{4} = \frac{6}{8}$$

estimate   A number close to an exact value. An estimate indicates *about* how much.

47 + 22 is about
50 + 20 or 70.

estimación   Número cercano a un valor exacto. Una estimación indica una cantidad approximada.

47 + 22 es aproximadamente
50 + 20 o 70.

expanded form/expanded notation   The representation of a number as a *sum* that shows the value of each digit.

536 is written as 500 + 30 + 6.

forma desarrollada/notación desarrollada   Representación de un número como la *suma* del valor de cada dígito.

536 puede escribirse como 500 + 30 + 6.

expression   A combination of numbers, *variables,* and at least one *operation.*

expresión   Combinación de números, *variables* y por lo menos una *operación.*

fact family   A group of related facts using the same numbers.

5 + 3 = 8	5 × 3 = 15
3 + 5 = 8	3 × 5 = 15
8 − 3 = 5	15 ÷ 3 = 5
8 − 5 = 3	15 ÷ 5 = 3

familia de operaciones   Grupo de operaciones relacionadas que usan los mismos números.

5 + 3 = 8	5 × 3 = 15
3 + 5 = 8	3 × 5 = 15
8 − 3 = 5	15 ÷ 3 = 5
8 − 5 = 3	15 ÷ 5 = 3

factor   A number that *divides* a whole number evenly. Also a number that is *multiplied* by another number.

factor   Número entre el que se *divide* otro número natural sin dejar residuo. También es cada uno de los números en una multiplicación.

factor pairs   The two factors that are multiplied to find a product.

pares de factores   Los dos factores que se multiplican para hallar un producto.

fluid ounce (fl oz)   A *customary* unit of *capacity*.

onza líquida (oz líq)   Unidad *usual* de *capacidad*.

foot (ft)   A *customary* unit for measuring *length*. Plural is feet.

1 foot = 12 inches

pie (pie)   Unidad *usual* de *longitud*.

1 pie = 12 pulgadas

formula   An *equation* that shows the relationship between two or more quantities.

fórmula   *Ecuación* que muestra la relación entre dos o más cantidades.

fraction   A number that represents part of a whole or part of a set.

$$\frac{1}{2}, \frac{1}{3}, \frac{1}{4}, \frac{3}{4}$$

fracción   Número que representa una parte de un todo o una parte de un conjunto.

$$\frac{1}{2}, \frac{1}{3}, \frac{1}{4}, \frac{3}{4}$$

frequency table   A table for organizing a set of *data* that shows the number of times each result has occurred.

tabla de frecuencias   Tabla para organizar un conjunto de *datos* que muestra el número de veces que se ha obtenido cada resultado.

Gg

gallon (gal)   A *customary* unit for measuring *capacity* for liquids.

1 gallon = 4 quarts

galón (gal)   Unidad *usual* de *capacidad* de líquidos.

1 galón = 4 cuartos

gram (g)   A *metric* unit for measuring *mass*.

gramo (g)   Unidad *métrica* de *masa*.

Gg

Greatest Common Factor (GCF)
The greatest of the common *factors* of
two or more numbers.

The greatest common factor of
12, 18, and 30 is 6.

máximo común divisor (M.C.D.)
El mayor de los *factores* comunes de dos
o más números.

El máximo común divisor de
12, 18 y 30 es 6.

Hh

hexagon   A *polygon* with six sides and
six *angles*.

hexágono   *Polígono* con seis lados
y seis *ángulos*.

hundredth   A *place-value* position. One
of one hundred equal parts.

In the number 0.05, 5 is in
the hundredths place.

centésima   *Valor posicional.*
Una de cien partes iguales.

En el número 4.57, 7 está en
el lugar de las centésimas.

Ii

Identity Property of Addition   For any
number, zero plus that number is the
number.

$$3 + 0 = 3 \text{ or } 0 + 3 = 3$$

propiedad de identidad de la suma
Para todo número, cero más ese número
da como resultado ese mismo número.

$$3 + 0 = 3 \text{ o } 0 + 3 = 3$$

Identity Property of Multiplication
If you *multiply* a number by 1, the
product is the same as the given
number.

$$8 \times 1 = 8 = 1 \times 8$$

**propiedad de identidad de la
multiplicación**   Si *multiplicas* un número
por 1, el *producto* es igual al número
dado.

$$8 \times 1 = 8 = 1 \times 8$$

improper fraction   A *fraction* with a
numerator that is greater than or equal to
the *denominator*.

$$\frac{17}{3} \text{ or } \frac{5}{5}$$

fracción impropia   *Fracción* con un
numerador igual al *denominador* o mayor
que él.

$$\frac{17}{3} \text{ o } \frac{5}{5}$$

input   A quantity that is changed to produce an output.

entrada   Cantidad que se modifica y produce un valor de salida.

intersecting lines   *Lines* that meet or cross at a point.

rectas secantes   *Rectas* que se intersecan o se cruzan en un punto común.

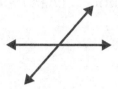

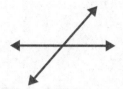

is equal to (=)   Having the same value. The (=) sign is used to show two numbers or *expressions* are equal.

es igual a (=)   Que tiene el mismo valor. Con el signo = se muestra que dos números o *expresiones* son iguales.

is greater than (>)   An inequality relationship showing that the number on the left of the symbol is greater than the number on the right.

$$5 > 3 \quad \text{5 is greater than 3.}$$

es mayor que (>)   Relación de desigualdad que muestra que el número a la izquierda del signo es más grande que el número a la derecha.

$$5 > 3 \quad \text{5 es mayor que 3.}$$

is less than (<)   An inequality relationship showing that the number on the left side of the symbol is less than the number on the right side.

$$4 < 7 \quad \text{4 is less than 7.}$$

es menor que (<)   Relación de desigualdad que muestra que el número a la izquierda del signo es más pequeño que el número a la derecha.

$$4 < 7 \quad \text{4 es menor que 7.}$$

isosceles triangle   A *triangle* with at least 2 sides of the same *length*.

triángulo isósceles   *Triángulo* que tiene por lo menos 2 lados del mismo *largo*.

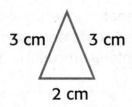

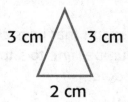

Kk

kilogram (kg)   A *metric* unit for measuring *mass*.

kilometer (km)   A *metric* unit for measuring *length*.

kilogramo (kg)   Unidad *métrica* de *masa*.

kilómetro (km)   Unidad *métrica* de *longitud*.

Ll

length   The measurement of a *line* between two points.

like fractions   *Fractions* that have the same *denominator*.

$$\frac{1}{5} \text{ and } \frac{2}{5}$$

line   A straight set of points that extend in opposite directions without ending.

line of symmetry   A *line* on which a figure can be folded so that its two halves match exactly.

line plot   A graph that uses columns of Xs above a *number line* to show frequency of *data*.

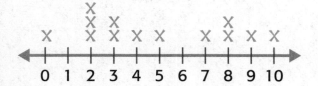

longitud   Medida de la distancia entre dos puntos.

fracciones semejantes   *Fracciones* que tienen el mismo *denominador*.

$$\frac{1}{5} \text{ y } \frac{2}{5}$$

recta   Conjunto de puntos alineados que se extiende sin fin en direcciones opuestas.

eje de simetría   *Recta* sobre la cual se puede doblar una figura de manera que sus mitades coincidan exactamente.

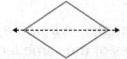

diagrama lineal   Gráfica que tiene columnas de X sobre una *recta numérica* para representar la frecuencia de los *datos*.

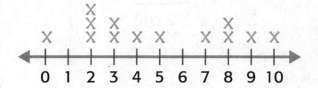

line segment A part of a *line* between two *endpoints*. The *length* of the line segment can be measured.

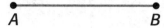

A   B

line symmetry A figure has *line symmetry* if it can be folded so that the two parts of the figure match, or are *congruent*.

liter (L) A *metric* unit for measuring *volume* or *capacity*.

1 liter = 1,000 milliliters

mass The amount of matter in an object. Two examples of units of measure would be gram and kilogram.

meter (m) A *metric* unit for measuring *length*.

metric system (SI) The decimal system of measurement. Includes units such as *meter*, *gram*, and *liter*.

mile (mi) A *customary* unit of measure for *length*.

1 mile = 5,280 feet

milliliter (mL) A *metric* unit for measuring *capacity*.

1,000 milliliters = 1 liter

millimeter (mm) A *metric* unit for measuring *length*.

1,000 millimeters = 1 meter

segmento de recta Parte de una *recta* entre dos *extremos*. La *longitud* de un segmento de recta se puede medir.

A   B

simetría axial Una figura tiene *simetria axial* si puede doblarse de modo que las dos partes de la figura coincidan de manera exacta.

litro (L) Unidad *métrica* de *volumen* o *capacidad*.

1 litro = 1,000 mililitros

masa Cantidad de materia en un cuerpo. El gramo y el kilogramo son dos ejemplos de unidades que se usan para medir la masa.

metro (m) Unidad *métrica* de *longitud*.

sistema métrico (SI) Sistema decimal de medidas que se basa en potencias de 10 y que incluye unidades como *el metro*, *el gramo* y *el litro*.

milla (mi) Unidad *usual* de *longitud*.

1 milla = 5,280 pies

mililitro (mL) Unidad *métrica* de *capacidad*.

1,000 mililitros = 1 litro

milímetro (mm) Unidad *métrica* de *longitud*.

1,000 milímetros = 1 metro

Mm

minuend The first number in a *subtraction* sentence from which a second number is to be subtracted.

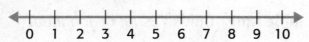

minuend **subtrahend** **difference**

mixed number A number that has a *whole number* part and a *fraction* part.

$$6\frac{3}{4}$$

multiple A multiple of a number is the *product* of that number and any whole number.

15 is a multiple of 5
because 3 × 5 = 15.

multiply (multiplication) An *operation* on two numbers to find their *product*. It can be thought of as repeated *addition*.

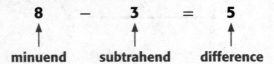

Nn

nonnumeric pattern Patterns that do not use numbers.

number line A *line* with numbers on it in order at regular intervals.

```
0  1  2  3  4  5  6  7  8  9  10
```

numerator The number above the bar in a *fraction*; the part of the fraction that tells how many of the equal parts are being used.

numeric pattern Patterns that use numbers.

---

minuendo El primer número en un enunciado de *resta* del cual se restará un segundo número

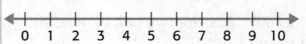

minuendo **sustraendo** **diferencia**

número mixto Número formado por un *número natural* y una parte *fraccionaria*.

$$6\frac{3}{4}$$

múltiplo Un múltiplo de un número es el *producto* de ese número y cualquier otro número natural.

15 es múltiplo de 5
porque 3 × 5 = 15.

multiplicar (multiplicación) *Operación* entre dos números para hallar su *producto*. También se puede interpretar como una *suma* repetida.

patrón no numérico Patrón que no usa números.

recta numérica *Recta* con números ordenados a intervalos regulares.

```
0  1  2  3  4  5  6  7  8  9  10
```

numerador El número que está encima de la barra de *fracción*; la parte de la fracción que indica cuántas de las partes iguales en que se divide el entero se están usando.

patrón numérico Patrón que usa números.

obtuse angle An *angle* that measures greater than 90° but less than 180°.

ángulo obtuso *Ángulo* que mide más de 90° pero menos de 180°.

obtuse triangle A *triangle* with one *obtuse angle*.

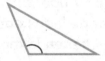

triángulo obtusángulo *Triángulo* con un *ángulo obtuso*.

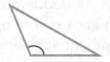

octagon A *polygon* with 8 sides and 8 *angles*.

octágono *Polígono* de 8 lados y 8 *ángulos*.

operation A mathematical process such as *addition* (+), *subtraction* (−), *multiplication* (×), or *division* (÷).

operación Proceso matemático como la *suma* (+), la *resta* (−), la *multiplicación* (×) o la *división* (÷).

order of operations Rules that tell what order to follow when evaluating an *expression*:
(1) Do the *operations* in *parentheses* first.
(2) *Multiply* and *divide* in order from left to right.
(3) *Add* and *subtract* in order from left to right.

orden de las operaciones Reglas que te indican qué orden seguir cuando evalúas una *expresión:*
(1) Resuelve primero las *operaciones* dentro de los *paréntesis*.
(2) *Multiplica* o *divide* en orden de izquierda a derecha.
(3) *Suma* o *resta* en orden de izquierda a derecha.

ounce (oz) A *customary* unit to measure *weight* or *capacity*.

onza (oz) Unidad *usual* de *peso* o *capacidad*.

output The result of an input quantity being changed.

salida Resultado que se obtiene al modificar un valor de entrada.

Pp

Pp

parallel lines   *Lines* that are the same distance apart. Parallel lines do not meet.

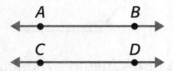

rectas paralelas   *Rectas* separadas por la misma distancia en cualquier punto. Las rectas paralelas no se intersecan.

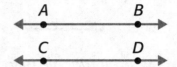

parallelogram   A *quadrilateral* with four sides in which each pair of opposite sides are *parallel* and equal in *length*.

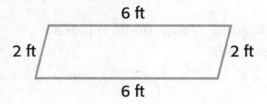

paralelogramo   *Cuadrilátero* en el que cada par de lados opuestos son *paralelos* y tienen la misma *longitud*.

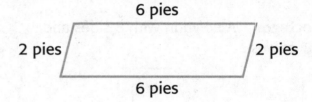

parentheses   The enclosing symbols (), which indicate that the terms within are a unit.

paréntesis   Los signos ( ) con que se encierran los términos, para indicar que cuando están adentro, forman una unidad.

partial products   A multiplication method in which the products of each place value are found separately, and then added together.

productos parciales   Método de multiplicación por el cual los productos de cada valor posicional se hallan por separado y luego se suman entre sí.

partial quotients   A dividing method in which the dividend is separated into sections that are easy to divide.

cocientes parciales   Método de división por el cual el dividendo se separa en secciones que son fáciles de dividir.

pattern   A sequence of numbers, figures, or symbols that follows a rule or design.

2, 4, 6, 8, 10

patrón   Sucesión de números, figuras o símbolos que sigue una regla o un diseño.

2, 4, 6, 8, 10

pentagon   A *polygon* with five *sides* and five *angles*.

pentágono   *Polígono* de cinco *lados* y cinco *ángulos*.

percent   A ratio that compares a number to 100.

porcentaje   Razón que compara un número con el 100.

perimeter   The distance around a shape or region.

perímetro   Distancia alrededor de una figura o región.

period   The name given to each group of three digits on a place-value chart.

período   Nombre dado a cada grupo de tres dígitos en una tabla de valor posicional.

perpendicular lines   *Lines* that meet or cross each other to form *right angles.*

rectas perpendiculares   *Rectas* que se intersecan o cruzan formando *ángulos rectos.*

pint (pt)   A *customary* unit for measuring *capacity*.

1 pint = 2 cups

pinta (pt)   Unidad *usual* de *capacidad.*

1 pinta = 2 tazas

place value   The value given to a *digit* by its position in a number.

valor posicional   Valor dado a un *dígito* según su posición en un número.

point   An exact location in space that is represented by a dot.

punto   Ubicación exacta en el espacio que se representa con una marca puntual.

polygon   A closed *plane figure* formed using *line segments* that meet only at their *endpoints.*

polígono   *Figura plana* cerrada formada por *segmentos de recta* que solo se unen en sus *extremos.*

pound (lb)   A *customary* unit to measure *weight* or *mass.*

1 pound = 16 ounces

libra (lb)   Unidad *usual* de *peso* o *masa.*

1 libra = 16 onzas

Pp

prime number A whole number with exactly two *factors*, 1 and itself.

7, 13, and 19

product The answer or result of a *multiplication* problem. It also refers to expressing a number as the *product* of its *factors*.

protractor An instrument used to measure angles.

número primo Número natural que tiene exactamente dos *factores*: 1 y sí mismo.

7, 13 y 19

producto Respuesta o resultado de un problema de *multiplicación*. Además, un número puede expresarse como el *producto* de sus *factores*.

transportador Instrumento con el que se miden los ángulos.

Qq

quadrilateral A shape that has 4 sides and 4 *angles*.

square, rectangle, and parallelogram

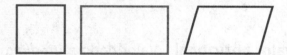

quart (qt) A *customary* unit for measuring *capacity*.

1 quart = 4 cups

quotient The result of a *division* problem.

cuadrilátero Figura que tiene 4 lados y 4 *ángulos*.

cuadrado, rectángulo y paralelogramo

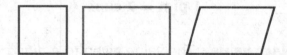

cuarto (ct) Unidad *usual* de *capacidad*.

1 cuarto = 4 tazas

cociente Respuesta o resultado de un problema de *división*.

Rr

ray A part of a *line* that has one *endpoint* and extends in one direction without ending.

semirrecta Parte de una *recta* que tiene un *extremo* y que se extiende sin fin en una dirección.

rectangle   A *quadrilateral* with four *right angles*; opposite sides are equal and *parallel*.

regroup   To use place value to exchange equal amounts when renaming a number.

remainder   The number that is left after one whole number is *divided* by another.

repeated subtraction   To subtract the same number over and over until you reach 0.

rhombus   A *parallelogram* with four *congruent* sides.

right angle   An *angle* with a measure of 90°.

right triangle   A *triangle* with one *right angle*.

rectángulo   *Cuadrilátero* con cuatro *ángulo rectos*; los lados opuestos son iguales y *paralelos*.

reagrupar   Usar el valor posicional para expresar una cantidad de otra manera.

residuo   Número que queda después de *dividir* un número natural entre otro.

resta repetida   Procedimiento por el que se resta un número una y otra vez hasta llegar a 0.

rombo   *Paralelogramo* con cuatro lados *congruentes*.

ángulo recto   *Ángulo* que mide 90°.

triángulo rectángulo   *Triángulo* con un *ángulo recto*.

round To change the value of a number to one that is easier to work with. To find the nearest value of a number based on a given *place value.*

redondear Cambiar el valor de un número a uno con el cual es más fácil trabajar. Hallar el valor más cercano a un número basándose en un *valor posicional* dado.

rule A statement that describes a relationship between numbers or objects.

regla Enunciado que describe una relación entre números u objetos.

Ss

second A unit of time.
60 seconds = 1 minute

segundo Unidad de tiempo.
60 segundos = 1 minuto

sequence The ordered arrangement of terms that make up a pattern.

secuencia Disposición ordenada de términos que forman un patrón.

simplest form A *fraction* in which the *numerator* and the *denominator* have no common *factor* greater than 1.
$\frac{3}{5}$ is the simplest form of $\frac{6}{10}$.

mínima expresión *Fracción* en la que el *numerador* y el *denominador* no tienen un *factor* común mayor que 1.
$\frac{3}{5}$ es la mínima expresión de $\frac{6}{10}$.

solve To replace a *variable* with a value that results in a true sentence.

resolver Despejar una *variable* y verdadera reemplazarla por un valor que haga que la ecuación sea.

square A *rectangle* with four *congruent* sides.

cuadrado *Rectángulo* de cuatro lados *congruentes.*

square unit A unit for measuring *area.*

unidad cuadrada Unidad para medir el *área.*

standard form/standard notation The usual way of writing a number that shows only its *digits,* no words.

537   89   1,642

forma estándar/notación estándar Manera habitual de escribir un número usando solo sus *dígitos,* sin usar palabras.

537   89   1,642

subtract (subtraction)   An *operation* on two numbers that tells the *difference*, when some or all are taken away. Subtraction is also used to compare two numbers.

$$14 - 8 = 6$$

subtrahend   A number that is *subtracted* from another number.

$$14 - 5 = 9$$
$$\uparrow$$
subtrahend

sum   The answer to an *addition* problem.

survey   A method of collecting *data.*

restar (resta)   *Operación* con dos números que indica la *diferencia,* entre ellos. Puede usarse para quitar una cantidad de otra o para comparar dos números.

$$14 - 8 = 6$$

sustraendo   Un número que se *resta* de otro número.

$$14 - 5 = 9$$
$$\uparrow$$
sustraendo

suma   Respuesta o resultado que se obtiene al sumar.

encuesta   Método para recopilar *datos.*

tally chart   A way to keep track of *data* using *tally marks* to record the number of responses or occurrences.

What is Your Favorite Color?	
Color	Tally
Blue	ⅢⅡ III
Green	IIII

tally mark(s)   A mark made to keep track of and display *data* recorded from a *survey.*

tenth   One of ten equal parts, or $\frac{1}{10}$.

term   Each number in a numeric pattern.

tabla de conteo   Manera de llevar la cuenta de los *datos* usando *marcas de conteo* para anotar el número de respuestas o sucesos.

¿Cuál es tu color favorito?	
Color	Conteo
azul	ⅢⅡ III
verde	IIII

marca de conteo   Marca que se hace para llevar un registro y representar *datos* recopilados en una *encuesta.*

décima   Una de diez partes iguales o $\frac{1}{10}$.

término   Cada número en un patrón numérico.

Tt

thousandth(s) One of a thousand equal parts, or $\frac{1}{1000}$. Also refers to a *place value* in a *decimal* number. In the *decimal* 0.789, the 9 is in the thousandths place.

milésima Una de mil partes iguales o $\frac{1}{1000}$. También se refiere a un *valor posicional* en un número *decimal*. En el *decimal* 0.789, el 9 está en el lugar de las milésimas.

three-dimensional figure A solid figure has three dimensions: *length,* width, and height.

figura tridimensional Figura sólida que tiene tres dimensiones: *largo,* ancho y alto.

ton (T) A *customary* unit to measure *weight.*

1 ton = 2,000 pounds

tonelada (T) Unidad *usual* de *peso.*

1 tonelada = 2,000 libras

trapezoid A *quadrilateral* with exactly one pair of *parallel* sides.

trapecio *Cuadrilátero* con exactamente un par de lados *paralelos.*

triangle A *polygon* with three sides and three *angles.*

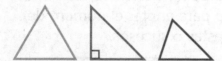

triángulo *Polígono* con tres lados y tres *ángulos.*

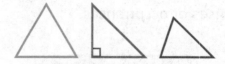

two-dimensional figure A figure that lies entirely within one plane.

figura bidimensional Figura que puede representarse en un plano.

Uu

unit square A square with a side length of one unit.

cuadrado unitario Cuadrado cuyos lados miden una unidad de longitud.

unknown The amount that has not been identified.

incógnita La cantidad que no ha sido identificada.

variable A letter or symbol used to represent an unknown quantity.

variable Letra o símbolo que se usa para representar una cantidad desconocida.

Venn diagram A diagram that uses *circles* to display elements of different sets. Overlapping *circles* show common elements.

diagrama de Venn Diagrama con *círculos* para mostrar elementos de diferentes conjuntos. Los *círculos* sobrepuestos indican elementos comunes.

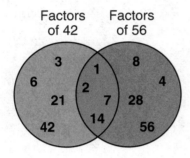

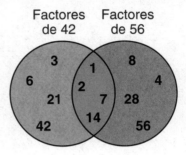

vertex The point where two *rays* meet in an *angle*.

vértice Punto donde se unen dos *semirrectas* formando un *ángulo*.

weight A measurement that tells how heavy an object is.

peso Medida que indica cuán pesado o liviano es un cuerpo.

word form/word notation The form of a number that uses written words.

forma verbal/notación verbal Forma de expresar un número usando palabras escritas.

yard (yd) A *customary* unit of *length* equal to 3 feet or 36 inches.

yarda (yd) Unidad *usual* de *longitud* igual a 3 pies o 36 pulgadas.

Zero Property of Multiplication
The property that states any number
multiplied by zero is zero.

$$0 \times 5 = 0 \qquad 5 \times 0 = 0$$

propiedad del cero de la multiplicación
Propiedad que establece que cualquier
número *multiplicado* por cero es igual
a cero.

$$0 \times 5 = 0 \qquad 5 \times 0 = 0$$

Name _____

Work Mat 1: Hundreds Chart

1	2	3	4	5	6	7	8	9	10
11	12	13	14	15	16	17	18	19	20
21	22	23	24	25	26	27	28	29	30
31	32	33	34	35	36	37	38	39	40
41	42	43	44	45	46	47	48	49	50
51	52	53	54	55	56	57	58	59	60
61	62	63	64	65	66	67	68	69	70
71	72	73	74	75	76	77	78	79	80
81	82	83	84	85	86	87	88	89	90
91	92	93	94	95	96	97	98	99	100

Work Mat 2: Thousands, Hundreds, Tens, and Ones Chart

Thousands	Hundreds	Tens	Ones

Name

Work Mat 3: Place-Value Chart

Ones			Thousands			Millions		
hundreds	tens	ones	hundreds	tens	ones	hundreds	tens	ones

Work Mat 4: Algebra Mat

$$=$$